Biochar - Applications in Agriculture and Environment

Edited by Kassio Ferreira Mendes and Rodrigo Nogueira de Sousa

Published in London, United Kingdom

Biochar - Applications in Agriculture and Environment
http://dx.doi.org/10.5772/intechopen.1008000
Edited by Kassio Ferreira Mendes and Rodrigo Nogueira de Sousa

Contributors
Adewale Nafiu, Adnan Akhter, Alex Coker, Dongjie Chen, Fude Liu, Giovani Rangel Malerba Alfonso, Godswill Aguiyi, Goldy De Bhowmick, Hongwen Sun, Hui-lian Xu, Jan Willem Straten, Jie Zhang, Jinyuan Liang, Kalidas Mainali, Kassio Ferreira Mendes, Kwasi Adusei-Fosu, Le Jiao, Mengmeng Kong, Peng Zhang, Rodrigo Nogueira de Sousa, Victor Chude, Xiaoyong Liu, Xiting Chen, Yangyang Xu, Yuning Wang, Muhammad Taqqi Abbas

First published in London, United Kingdom, 2026 by IntechOpen
IntechOpen is the global imprint of INTECHOPEN LIMITED, registered in England and Wales, registration number: 11086078, 167-169 Great Portland Street, London, W1W 5PF, United Kingdom

For EU product safety concerns: IN TECH d.o.o., Prolaz Marije Krucifikse Kozulić 3, 51000 Rijeka, Croatia, info@intechopen.com or visit our website at intechopen.com.

British Library Cataloguing-in-Publication Data
A catalogue record for this book is available from the British Library

Biochar - Applications in Agriculture and Environment
Edited by Kassio Ferreira Mendes and Rodrigo Nogueira de Sousa
p. cm.
Print ISBN 978-1-83634-362-2
Online ISBN 978-1-83634-361-5
eBook (PDF) ISBN 978-1-83634-363-9

Meet the editors

Kassio Ferreira Mendes is an Associate Professor of Traceability in Agricultural and Environmental Quality at the Center for Nuclear Energy in Agriculture (CENA), University of São Paulo (USP), Brazil. He is an accredited advisor in the Interunit Graduate Program in Applied Ecology (ESALQ/CENA) and in the Graduate Program in Sciences (Nuclear Energy in Agriculture) at CENA. He earned his Ph.D. and completed a postdoctoral fellowship in Nuclear Energy in Agriculture at USP, including a research exchange at the University of Minnesota, USA. His expertise includes integrated weed management, environmental behavior of herbicides, biochar application in agriculture and the environment, and pesticide traceability using radioisotopes and chromatographic techniques.

Rodrigo Nogueira de Sousa holds a Ph.D. in Soils and Plant Nutrition from the Luiz de Queiroz College of Agriculture (ESALQ), University of São Paulo (USP), Brazil. He obtained his bachelor's degree in Agronomy Engineering and a master's degree in Soils and Plant Nutrition from the Federal University of Viçosa, Brazil. During his academic training, he conducted research on soil organic matter and fertility, including scientific initiation at the Stable Isotope Laboratory at UFV. He participated in international academic programs at North Carolina A&T State University and North Carolina State University in the USA. His expertise lies in soil fertility, soil chemistry, and plant nutrition. He is currently doing postdoctoral research at the Center for Nuclear Energy in Agriculture (CENA/USP).

Contents

Preface

Biochar has rapidly evolved from a traditional soil amendment into a multifunctional material with broad relevance across agriculture, environmental science, and sustainable resource management. Produced through the thermochemical conversion of biomass under oxygen-limited conditions, biochar exhibits a unique combination of physicochemical stability, surface reactivity, and porosity that underpins its diverse applications. As global challenges related to soil degradation, food security, climate change, and environmental contamination intensify, biochar-based technologies have gained increasing attention as part of integrated, low-carbon solutions for agroecosystem management.

The edited volume *Biochar – Applications in Agriculture and Environment* brings together recent advances that span fundamental processes, technological innovations, and applied case studies. The chapters collectively reflect the transition of biochar research from empirical observations toward mechanistic understanding, systems integration, and field-level implementation. The volume is organized into two complementary sections that guide the reader from production and functionalization strategies to agricultural, environmental, and plant health applications.

The first section focuses on biochar production pathways, functional modification, and advanced analytical and technological approaches. The opening chapter, "Biochar Synergy with Smart Agriculture and Environment: From Soil Amendment to Precision Regulation Systems", introduces biochar as an enabling component of smart and climate-resilient agricultural systems. It highlights the integration of biochar with sensors, nanotechnology, artificial intelligence, and closed-loop management of carbon, water, and nutrients. This systems-level perspective sets the conceptual foundation for the volume.

The following chapter, "Combined Thermochemical Conversion as a Pretreatment of Manure Waste for Environmental Applications: An Overview", examines biochar production in the context of manure management and anaerobic digestion. It emphasizes thermochemical pretreatment strategies that enhance biogas yields while producing nutrient-enriched biochar, thereby contributing to waste valorization and environmental sustainability.

In "Recent Progress in the Influence of B- and N-Doped Primary and Secondary Hydrochar-Based Electrodes from Renewable Raw Materials on the Capacitive Performance for Energy Storage Applications" attention shifts to material functionalization and emerging non-agricultural uses of biochar-derived carbons. This chapter reviews the role of heteroatom doping and hydrothermal carbonization in tailoring carbon materials for energy storage, illustrating the expanding technological scope of biochar research.

The first section concludes with "Isotopic Tracers and Biochar in Soil Fertility and Plant Nutrition: Mechanistic Insights into Nutrient Cycling, Carbon Dynamics and Agroecosystem Functioning – A Review". This chapter addresses a critical methodological gap by demonstrating how stable and radioactive isotopes can disentangle biochar-driven processes in soil–plant systems. By resolving carbon turnover, nutrient fluxes, and priming effects, isotopic approaches provide a robust framework for advancing mechanistic understanding and improving agronomic recommendations.

The second section centers on agricultural and environmental applications, with an emphasis on soil health, crop productivity, and plant protection. "Research and Application Progress of Biochar in Amelioration of Saline-Alkali Soil" synthesizes current knowledge on the role of biochar in improving saline–alkali soils, detailing its effects on soil physicochemical properties, nutrient availability, microbial activity, and carbon sequestration. The chapter highlights both opportunities and limitations associated with large-scale implementation.

Field-based evidence is further explored in "Use of Biochar and Compost in Soil Remediation, Growth and Yield of Cereals and Legume Crops in Niger State, Nigeria". This chapter presents applied research demonstrating how biochar, alone or in combination with compost, can remediate problem soils, enhance crop growth, and improve yields under smallholder farming conditions, thereby linking biochar technology to food security and rural livelihoods.

The volume concludes with "Biochar as a Management Tool for Soilborne *Fusarium* and *Phytophthora* Pathogens", which expands the discussion to plant health and disease suppression. By focusing on soilborne pathogens of major economic importance, this chapter highlights biochar's indirect mechanisms of action through rhizosphere modification, microbial interactions, and soil environmental changes, positioning biochar as a complementary tool in sustainable plant disease management.

Together, the chapters in this edited volume provide a coherent overview of biochar research across scales, from molecular-level mechanisms and material engineering to field applications and agroecosystem functioning. By integrating production technologies, mechanistic insights, and practical applications, the book aims to support researchers, practitioners, and policymakers seeking science-based strategies for sustainable agriculture and environmental management.

We editors gratefully acknowledge all contributing authors for their expertise and commitment, as well as the reviewers whose constructive comments strengthened the quality of this volume. Appreciation is also extended to the IntechOpen editorial team for their professional support throughout the publication process.

Dr. Kassio Ferreira Mendes and Dr. Rodrigo Nogueira de Sousa
Professor,
Center for Nuclear Energy of Agriculture,
University of São Paulo,
Piracicaba, SP, Brazil

Section 1

Biochar Production, Functionalization and Advanced Environmental Technologies

Chapter 1

Biochar Synergy with Smart Agriculture and Environment: From Soil Amendment to Precision Regulation Systems

Mengmeng Kong, Xiaoyong Liu, Dongjie Chen, Yangyang Xu, Jie Zhang, Xiting Chen, Jinyuan Liang and Hui-lian Xu

Abstract

Biochar, a carbon-rich product of biomass pyrolysis, is undergoing a paradigm shift from a conventional soil amendment to a precision regulation tool within smart agricultural systems. This review examines the evolving applications of biochar, integrating nanotechnology, sensor systems, and artificial intelligence (AI) to facilitate precision management of carbon, water, and nutrient cycles. We investigate how the combination of biochar and intelligent technologies fosters closed-loop systems that enhance resource efficiency, mitigate climate change, and promote environmental sustainability. Key topics include the functional enhancement of biochar via nanosensor integration and its role in climate-smart agriculture. By synthesizing scientific evidence and case studies, we highlight biochar's potential to tackle global agricultural challenges while underscoring the necessity for collaborative governance and technological innovation.

Keywords: biosensor, nanotechnology, closed-loop management, climate-smart agriculture (CSA), precision environmental management

1. Introduction

Biochar's legacy traces back to the ancient Amazonian Terra Preta de Índio (Indian Black Soil), where its use transformed infertile tropical soils into enduringly fertile ecosystems, demonstrating prehistoric mastery of soil enhancement through carbon-rich amendments [1]. Modern science confirms biochar's multifaceted value: its porous structure (300–2000 m^2/g surface area) improves water retention by up to 39% and soil porosity by 15–35% [2–4]; oxygen-rich surface groups (carboxyl, hydroxyl) enhance cation exchange capacity (CEC), allowing for better retention of nutrients like NH_4^+ and K^+ [5, 6]; its alkalinity counters soil acidity [7]; and it sustains microbial habitats while sequestering carbon for centuries [1, 8–12].

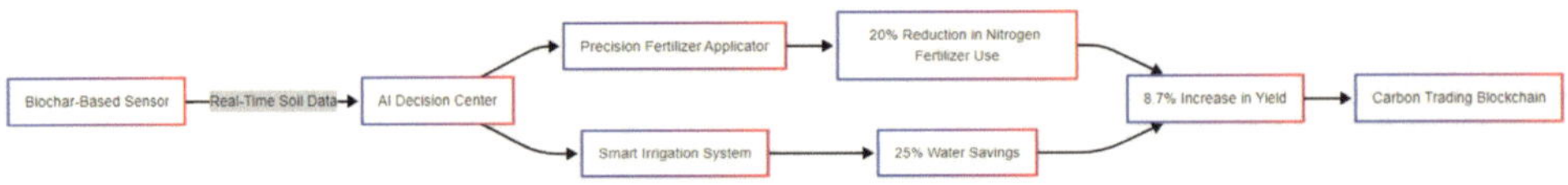

Figure 1.
The operational framework of the AI-biochar system in Jiangsu Province, China.

However, traditional applications face critical constraints: a passive "input-only" role, reliance on empirical field trials, narrow system boundaries, and non-optimized "one-size-fits-all" approaches [13]. These limitations collide with today's agricultural challenges, including the need to increase output by 60% by 2050 amid shrinking arable land [14], agriculture's 17–21% contribution to greenhouse gas emissions [12, 15], water scarcity (accounting for 70% of global usage), soil degradation, and pollution from agrochemicals [12, 14].

Intelligent technologies are now catalyzing a paradigm shift, transforming biochar from a basic amendment into a dynamic smart agriculture regulator [15]. This evolution hinges on three advances:

1. Functional innovation, where biochar becomes an "intelligent carrier" embedded with nanosensors (optical, electrochemical) or stimuli-responsive nanomaterials (e.g., pH/temperature-sensitive hydrogels) [16–18]. These engineered composites enable real-time soil monitoring (moisture, nutrients, pollutants) and adaptive responses—moving beyond static amendment to active "sense-and-respond" systems [19–25].

2. System integration, repositioning biochar as a hub for closed-loop resource management. It connects carbon loops (12–50% CO_2-equivalent reduction [15, 26]), water cycles (precision irrigation via enhanced retention [8]), and nutrient recycling (slow-release fertilizers from waste-derived biochar [2, 27, 28]), aligning with Climate-Smart Agriculture (CSA) goals [29].

3. AI-driven decision-making, where biochar-based sensors feed IoT networks. Machine learning algorithms then optimize variable-rate applications—reducing fertilizer use by 30% while maintaining yields [15, 30, 31].

This triad of advances converges in real-world systems like Jiangsu Province's AI-biochar project for rice-wheat rotations [32]. By integrating real-time soil and weather monitoring, AI-prescribed nitrogen management, and biochar's nutrient-retention capabilities, the project reduced fertilizer inputs by 20%, boosted yields by 8.7%, and curtailed N_2O emissions—showcasing scalable synergy between ecological function and digital precision (**Figure 1**).

Thus, biochar's technological metamorphosis—from ancestral soil enhancer to AI-augmented systems regulator—offers a pathway to reconcile food security, resource efficiency, and environmental resilience in the Anthropocene.

2. Biochar as an intelligent carrier: Nanotechnology and sensor integration

In the context of smart agriculture, biochar has emerged as an intelligent core carrier endowed with sensing, response, and regulatory capabilities through the application of nanotechnology and sensor integration. This evolution provides innovative tools for precision environmental management [33–36].

2.1 Precision regulation of biochar structure driven by nanotechnology

To optimize its performance as an intelligent carrier (loading capacity, targeting, response sensitivity), nano-scale structural design of biochar is imperative (**Table 1**).

2.1.1 Directed optimization of pore structure

Pyrolysis condition control: The temperature (300–700°C), heating rate, and residence time significantly influence pore development. Elevated temperatures (>500°C) facilitate micropore formation, while lower temperatures (300–500°C) help retain oxygen-containing functional groups [36].

Precursor selection: Lignocellulosic materials (such as wood and straw) yield abundant pores, whereas manured biochar tends to have a higher ash content and more complex pore structures [13].

Physical/chemical activation: The activation process using steam or $KOH/ZnCl_2$ expands pores and increases specific surface area (up to >3000 m^2/g), thus enhancing loading capacity [38].

2.1.2 Surface functionalization modification

a. Acid/base modification: Oxidation with HNO_3/H_2SO_4 increases carboxyl and hydroxyl groups; treatment with NaOH/KOH enhances negative charge and hydroxyl content [38, 39].

b. Metal (hydroxide) loading: Techniques such as impregnation-calcination or coprecipitation allow for the loading of Fe_3O_4, MgO, $Al(OH)_3$, and $La(OH)_3$, facilitating magnetic separation or selective phosphorus adsorption [40, 41].

Regulation strategies	Specific methods	Main objectives/effects	Key performance improvements
Pore optimization	High-temperature pyrolysis, steam activation, chemical activation (KOH, $ZnCl_2$)	Increase specific surface area, regulate pore size distribution (micro-/mesoporous ratio)	Loading capacity ↑, diffusion rate ↑, mass transfer efficiency ↑
Surface functionalization	Acid/base treatment, metal (hydroxide) loading, organic molecule grafting, heteroatom doping	Introduce/enhance specific functional groups (–COOH, –OH, $–NH_2$, quaternary ammonium, etc.)	Adsorption selectivity ↑, catalytic activity ↑, wettability control, electrochemical performance ↑
Nano-composite	Loading nZVI, composite with carbon nanotubes/graphene, embedding in hydrogels, composite with clay	Synergize multiple material advantages, construct multifunctional interfaces	Multifunctionality (adsorption + catalysis+ magnetic recovery + conductivity + water retention), stability ↑, responsiveness ↑

Table 1.
Main methods and effects of nanotechnology on biochar carrier performance [37].

c. Organic molecule grafting: The grafting of quaternary ammonium salts, amino groups, or hydrophobic groups modifies cation exchange capacity, metal chelation, or wettability [38, 39].

d. Heteroatom doping: The introduction of nitrogen, sulfur, or phosphorus alters the surface electronic structure, enhancing catalytic or adsorption properties [39, 42].

2.1.3 Advanced nano-composite strategies

a. Biochar-based nano-zero-valent iron (nZVI-BC): Biochar stabilizes nZVI nanoparticles, synergistically enhancing the reduction of heavy metals and organic pollutants [43].

b. Biochar-carbon nanotube/graphene composites: π-π stacking facilitates the formation of three-dimensional conductive networks, thereby improving electrochemical performance [44].

c. Biochar-clay mineral composites: The combination of adsorption and ion exchange capabilities enhances the removal of nutrients and heavy metals [45].

d. Biochar-hydrogel composites: The integration of biochar within hydrogels merges adsorption with water retention, offering intelligent management of water and fertilizers [46].

2.2 Principles, construction, and integration of biochar-based sensors

Functionalized nano-biochar, characterized by its unique physico-chemical properties—including conductivity, catalytic activity, large surface area, and abundant adsorption sites—serves as an ideal material for the development of high-performance, cost-effective sensors for in-situ agricultural environmental monitoring. The primary types of these sensors are outlined below (**Table 2**).

2.3 Application cases

See **Table 3**.

2.4 Challenges and future directions

a. Stability and reliability of sensing performance: Complex agricultural environments (soil texture, organic matter, pH, microbial activity) may interfere with sensor performance. There is a need for the development of robust sensing interfaces, anti-pollution/self-cleaning materials, and in-situ calibration methods.

b. Long-term stability and service life: Ensuring long-term (months to years) field stability of biochar-based sensors remains a challenge, necessitating research into material aging mechanisms and self-repairing systems.

c. Scalable production and cost control: Simplification and standardization of functionalized nano-biochar and sensor fabrication are essential for cost reduction and practical implementation.

DOI: http://dx.doi.org/10.5772/intechopen.1011847

Sensor type	Core principle	Typical detection targets	Role of biochar	Agricultural environmental application examples
Electrochemical sensors [47]	Current/potential/impedance changes	NO_3^-, NH_4^+, PO_4^{3-}, K^+, H^+, heavy metals (Cd^{2+}, Pb^{2+}), pesticides	Acts as an electrode modification material, providing a conductive substrate, (electro) catalytic activity, and high adsorption enrichment	Real-time monitoring of soil/irrigation water nutrients, rapid pollutant screening
Fluorescent sensors [48]	Fluorescence intensity change/wavelength shift	Heavy metals (Hg^{2+}, Cu^{2}), pesticides, H_2O_2 (stress marker)	Functions as a fluorescent carrier/quencher or participates in recognition reactions	Detection of trace pollutants in water/soil, early diagnosis of plant stress
Colorimetric sensors [49]	Solution/test strip color change	pH, heavy metals, pesticides, H_2O_2, glucose	Serves as a color reaction catalyst (nanozyme), color substance carrier, and adsorption enhancer	On-site rapid visual detection using test strips/microfluidics
Resistance/capacitance sensors [50]	Resistance/capacitance change with environmental parameters (humidity, gas)	Soil volumetric water content (VWC), NH_3, H_2S	Sensitive material, with dielectric/resistive properties that change with the environment	Precision irrigation control, environmental monitoring in livestock sheds, soil aeration assessment
Integrated sense-release system [51–53]	Sensing signal triggers carrier release/activation	Nutrient concentration thresholds, rhizosphere signaling molecules, pollutant levels	Combines sensing and responsive release functions	On-demand precision fertilization/pesticide application, *in situ* soil remediation, intelligent rhizosphere regulation

Table 2.
Main types of biochar-based sensors, principles, and agricultural environmental applications [47–53].

Case scenario	Integration of decision-making and execution	Advantages
In situ monitoring and simultaneous remediation of farmland heavy metal pollution [54]	Electrochemical or colorimetric sensor networks utilizing nZVI-BC composites monitor dynamic changes in Cd, As, etc., in soil pore water/leachate. When sensor-detected heavy metal concentrations exceed safety thresholds, the system automatically alarms and triggers enhanced adsorption/reduction of heavy metals through nZVI-BC repair materials deployed in the same area.	Integrated monitoring and remediation, dynamic evaluation of repair effects, precise and on-demand repair, and cost reduction.
Closed-loop system for precision water-fertilizer management [51, 52, 55]	Distributed biochar-based electrochemical/capacitive sensor networks monitor NO_3^-, NH_4^+, K^+, PO_4^{3-} content and volumetric water content (VWC) at various soil depths. Monitoring data is transmitted *via* IoT to cloud AI platforms, which generate optimal water-fertilizer decisions (irrigation volume, fertilization formula, timing) by integrating crop models, growth stages, and weather forecasts. Continuous monitoring of soil conditions following water-fertilizer application forms a closed-loop feedback system, continuously optimizing decisions.	Enhances efficiency in water-fertilizer management, leading to improved crop yield and resource utilization.
Intelligent control of protected agricultural environment [56]	Biochar-based sensors integrated into greenhouses or plant factories monitor air temperature/humidity, CO_2 concentration, substrate nutrients/water, and potential NH_3/H_2S gases. Data-driven environmental control systems (ventilation, humidification, CO_2 supplementation, irrigation fertilization) operate automatically to maintain optimal crop growth conditions.	Creates highly controllable environments, maximizes resource use efficiency, and achieves consistent high-quality, standardized production.

Table 3.
Implementation scenarios in smart agriculture and environmental monitoring [51, 52, 54–56].

d. Multi-parameter integration and selectivity enhancement: The development of miniaturized, integrated biochar sensor probes for simultaneous in-situ monitoring of multiple parameters (N, P, K, pH, moisture, temperature, heavy metals) is critical.

e. AI-driven intelligent analysis and prediction: Deep integration of sensor data with AI (machine learning, deep learning) is essential for data denoising, fault diagnosis, and intelligent decision optimization.

f. Standardization and regulations: The absence of performance evaluation standards and environmental safety assessment frameworks for biochar-based sensors necessitates immediate attention.

2.5 Conclusion

The convergence of nanotechnology and sensor technology has transformed the role of biochar in smart agriculture. By precisely regulating nanostructure and surface chemistry, and incorporating sensing functionalities, biochar has transitioned from a passive amendment to an intelligent carrier equipped with environmental perception, information transmission, and responsive capabilities. Biochar-based sensors facilitate in-situ, real-time, and cost-effective monitoring of critical parameters within soil-water-crop systems, while the "Sense-Release" integration establishes closed-loop management paradigms. Despite existing challenges related to long-term stability and scalability, ongoing advancements in materials science, the Internet of Things (IoT), and artificial intelligence (AI) will enhance biochar's position as a fundamental intelligent carrier in smart agriculture, promoting its digital, intelligent, and ecological transformation.

3. Biochar-driven "carbon-water-nutrient" closed-loop management system: Establishing the resource circulation core of smart agriculture

The fundamental value of biochar in smart agriculture resides in its capacity to serve as a nexus for carbon, water, and nutrient cycles, facilitating the transition from traditional linear consumption models to efficient, collaborative closed-loop management systems. This section explores how biochar reconfigures resource circulation pathways and how advanced technologies enable precise sensing, informed decision-making, and dynamic regulation to achieve synergistic objectives of resource efficiency, emission reduction, and productivity enhancement.

3.1 Biochar: The multifunctional "engine" of closed-loop systems

Biochar significantly impacts the migration, transformation, and utilization efficiency of carbon, water, and nutrients within soil-plant systems, providing the material foundation for closed-loop management.

3.1.1 Carbon cycle reconstruction

See **Table 4**.

Regulatory methods	Mechanisms: enhancing solid retention from source to sink	
Long-term sequestration in inert carbon pools	Biochar's highly aromatic and biodegradation-resistant organic carbon (50–90%) remains stable in soils for centuries to millennia, thereby mitigating immediate CO_2 emissions from biomass decomposition. Global models indicate that sustainable biochar application could sequester up to 1 billion tons of CO_2 equivalent annually [29, 57].	
Bidirectional regulation of the priming effect	Initial biochar application may temporarily "prime" the decomposition of native soil organic matter (SOM), but high-quality biochar—particularly that produced at high temperatures—suppresses SOM mineralization through structural stability and adsorption protection, indirectly enhancing overall soil carbon storage [9, 58].	
Synergistic greenhouse gas emission reduction	N_2O emission reduction	Biochar can decrease N_2O emissions by 30–70% through substrate adsorption, enhanced aeration, and promotion of N_2O-reducing microbial genes [59, 60].
	CH_4 flux regulation	In paddy fields, biochar has been shown to reduce CH_4 emissions by fostering CH_4-oxidizing bacteria, while its impact in upland areas is negligible [61].

Table 4.
Mechanisms for carbon cycle reconstruction [9, 29, 57–61].

3.1.2 Water cycle optimization: From water retention to efficiency enhancement

See **Table 5**.

Regulatory methods	Mechanisms: From water retention to efficiency enhancement
Physical structure improvement and water retention enhancement	Biochar's porous structure lowers soil bulk density, increases total porosity, and enhances water retention (with field water capacity often rising by 15–50%), thereby reducing evaporation and surface runoff [62, 63]. In the semi-arid Loess Plateau of China, biochar application boosted soil water retention by 25%, alleviating water stress and achieving a 25% reduction in water usage [64].
Enhanced plant water use efficiency (WUE)	Improved soil moisture encourages root development and regulates stomatal conductance, resulting in increased dry matter production per unit of water used [65].

Table 5.
Mechanisms for water cycle optimization [62–65].

3.1.3 Nutrient cycle efficiency: From loss to recovery

See **Table 6**.

Regulatory methods	Mechanisms: From loss to recovery
Expanded nutrient storage and adsorptive slow release	Biochar's extensive surface area and functional groups effectively adsorb NH_4^+, NO_3^-, PO_4^{3-}, and K^+, reducing leaching (20–80% for nitrate) and facilitating slow nutrient release, thereby extending fertilizer efficiency [66, 67].
Regulation of nutrient	Transformation process nitrogen (N) transformation: The adsorption of NH_4^+ and the inhibition of nitrification decrease NO_3^- accumulation and N_2O emissions [68].
	Phosphorus (P) activation: Biochar raises pH in acidic soils, diminishing Fe/Al phosphate fixation, and releases fixed P by adsorbing Al^{3+}/Fe^{3+} [69, 70].
	The availability of potassium (K) and trace elements: Biochar is inherently rich in potassium, particularly when derived from straw-based raw materials. Its cation exchange capacity (CEC) enhances with aging, facilitating the retention and exchange of cations such as K^+, Ca^{2+}, and Mg^{2+}. Additionally, the surface functional groups of biochar have the capacity to complex trace metal elements, thereby influencing their availability and mobility.
Accelerated microbe-driven nutrient turnover	Biochar's porous structure fosters microbial biomass and activity, expediting the mineralization of organic matter and the release of nutrients [33].

Table 6.
Mechanisms for nutrient cycle efficiency [33, 66–70].

3.2 Empowerment through intelligent technologies: Sensing, decision-making, and execution in closed-loop systems

While biochar provides the essential physical framework for closed-loop management, intelligent technologies—such as AI, IoT, and big data—equip the system with "nerves," "brain," and "limbs," enabling a transition from passive responses to proactive precision regulation.

3.2.1 Intelligent sensing layer

See **Table 7**.

Multi-dimension		**Real-time data collection**
In situ sensing networks	Dense deployment of low-cost, low-power smart sensor nodes	Soil sensors: Monitor volumetric water content (VWC), matrix potential, temperature, electrical conductivity (EC), pH, and ion-selective electrodes for nitrates (NO_3^-), ammonium (NH_4^+), and potassium (K^+) [47].
		Environmental sensors: Include weather stations and groundwater quality monitors.
		Crop physiological sensors: Utilize stem flow meters, leaf temperature sensors, and canopy hyperspectral imaging.
Biochar status monitoring		Embed fluorescent tracers or use wireless sensors (NFC/RFID) to track biochar distribution and aging [37].
Space-air-ground integrated monitoring		Combine UAV remote sensing with satellite data for macro-level insights.

Table 7.
Multidimensional, real-time data collection in the intelligent sensing layer [37, 47].

3.2.2 Intelligent decision-making layer

See **Table 8**.

Building of models	**AI-driven model prediction and optimization**
Multi-source data fusion and feature extraction	AI (deep learning) processes heterogeneous spatio-temporal data to extract key features such as water stress and nitrogen nutrition indices [71].
Mechanistic-AI hybrid model construction	Carbon cycle simulation: A modified RothC model, combined with real-time data, predicts soil organic carbon dynamics and greenhouse gas fluxes [71].
	Water-nutrient-crop coupling models: Integrate soil water dynamics (HYDRUS), nutrient transport (NLEAP), and crop growth models (WOFOST), utilizing AI for parameter calibration [72].
Closed-loop optimization decision engine	Employs reinforcement learning or multi-objective optimization to integrate real-time data, predictive outputs, and management goals, generating optimal regulatory instructions [72].

Table 8.
AI-driven model prediction and optimization in the intelligent decision-making layer [71, 72].

3.2.3 Intelligent execution layer

See **Table 9**.

3.3 Closed-loop efficiency: Typical cases and quantitative evidence

See **Table 10**.

Regulatory methods	Precision response and dynamic regulation
Water-char collaborative intelligent irrigation	AI-driven variable rate irrigation (VRI) adjusts water supply based on biochar-modified soil water retention maps and real-time demand predictions [73].
Fertilizer-char collaborative intelligent fertilization	Base fertilizer optimization: Adjusts dosage based on soil nutrient levels and biochar's adsorption potential.
	Topdressing precision: Controls the timing and type of topdressing based on real-time nutrient monitoring and biochar release dynamics [73].
Biochar "activity" regulation	Investigates external stimuli (electric/magnetic fields) or responsive materials to trigger biochar's nutrient release or pollutant adsorption on demand [74].

Table 9.
Precision response and dynamic regulation in the intelligent execution layer [73, 74].

Cases		**China's loess plateau: Water-char-nutrient synergy in dryland agriculture** [75–78].	**Netherlands greenhouse** Horticulture: AI-enabled Carbon-water-nutrient closed-loop Precision control [72, 79–81].
Challenges		Severe soil erosion, drought, soil infertility, and low fertilizer efficiency.	High resource consumption (water, fertilizer, energy), nutrient leaching, and greenhouse gas emissions in high-input systems.
Solutions		Large-scale application of straw biochar (20–40 t/ha) along with soil and water conservation measures; implementation of soil moisture sensor networks for supplementary irrigation; use of biochar-based slow-release fertilizers.	Mixing highly adsorptive biochar (5–15% v/v) into soilless culture substrates; establishing dense sensor networks; utilizing AI models to optimize nutrient solutions, irrigation, and CO_2 enrichment [72, 79, 81].。
Closed-loop efficiency	Water	Soil water retention increased by 18–35%, and water use efficiency improved by 20–30% [75]. Biochar enhanced substrate water retention, with AI control achieving 90% water-fertilizer use efficiency and 30–40% water savings [72, 79, 80].	
	Carbon	Soil organic carbon (SOC) increased, with carbon sequestration rates of 1.5–3.0 t C/ha/yr., and N_2O emissions reduced by 35–60% [75, 76].	Biochar sequestered organic carbon, and precise CO_2 control maximized photosynthetic efficiency, reducing fossil fuel-derived CO_2 emissions [72].
	Nutrients	Nitrogen use efficiency improved by 15–25%, phosphorus availability increased by 20–40%, and crop yields rose by 10–25% [75, 77].	Biochar reduced nutrient loss, with AI control achieving 25–35% fertilizer savings and a 70% reduction in nitrate leaching [71, 79, 80].
Intelligent upgrades/ outputs		UAV-based NDVI mapping for variable topdressing, optimizing resource precision [75, 78].	Achieved stable yields, improved quality (sugar content, vitamin C), and reduced environmental footprint [80].

Table 10.
Typical cases and quantitative evidence [71, 72, 75–81].

3.4 Challenges and advancing directions for closed-loop systems

a. Spatio-temporal heterogeneity and prediction challenges of biochar effects: Long-term field experiments and advanced data mining techniques are essential for developing precise predictive models.

b. Sensor cost, reliability, and deployment density: Innovations in high-precision, low-cost in-situ sensors and optimized deployment strategies are crucial.

c. Model complexity and edge intelligence: The development of lightweight AI models (TinyML) is necessary for facilitating real-time closed-loop responses.

d. System integration and standardization: Achieving interoperability among sensors and platforms requires the establishment of open standards for data interfaces and communication protocols.

e. Economic feasibility and scalable promotion: Policy support and cost reduction through economies of scale are vital for quantifying comprehensive benefits, including yield increases, cost savings, and emission reductions.

3.5 Conclusion

The biochar-driven "carbon-water-nutrient" closed-loop management system represents a significant innovation in smart agricultural resource management. Serving as the physical core, biochar enhances resource circulation through carbon sequestration, water conservation, and nutrient retention, while intelligent technologies function as the neural center, enabling precise collaborative regulation. Global case studies illustrate its potential to enhance resource efficiency (over 25% water savings and over 20% fertilizer savings), reduce environmental pollution (30–70% emissions reduction), and ensure food security (10–25% yield increase). Despite challenges related to prediction accuracy and system integration, the closed-loop system is poised to become a foundational element of smart agriculture, driving sustainable global agricultural transformation. Future research should prioritize the deep integration of biochar, intelligent technologies, and management practices to facilitate large-scale application.

4. Role of biochar in reconstructing climate-smart agriculture (CSA) systems

Climate-Smart Agriculture (CSA) seeks to achieve three interrelated objectives: sustainably increasing agricultural productivity, enhancing resilience to climate change, and minimizing greenhouse gas emissions or enhancing carbon sinks. Biochar, with its distinctive physico-chemical properties and multifaceted impacts on soil-plant-atmosphere systems, is transitioning from a passive soil amendment to a critical engineering tool for actively reconstructing and optimizing CSA systems, offering innovative and comprehensive solutions to meet these three objectives.

4.1 Core mechanisms of biochar in empowering CSA: A multilevel action framework

Biochar's contribution to CSA is multifaceted, forming a synergistic framework through its effects on soil physical, chemical, and biological processes, as well as interfacial reactions (**Figure 2**):

4.2 Cornerstone role: A significant carbon sink and emission reduction engine

4.2.1 Long-term carbon sequestration

The chemical stability of biochar is attributed to its highly aromatic carbon structure, which has a mean residence time (MRT) in soils ranging from centuries to millennia, significantly surpassing that of uncharred organic matter (which lasts only years to decades). Global assessments by Lehmann et al. [82] indicate that biochar's carbon fixation efficiency exceeds 50%, markedly higher than composting (<20%). Large-scale models [29] estimate that the global annual sequestration potential of biochar is approximately 1 billion tons of CO_2 equivalent, positioning it as one of the most promising negative emission technologies (NETs) in agriculture.

4.2.2 Greenhouse gas (GHG) emission reduction

a. N_2O emission reduction: Biochar significantly contributes to reducing soil N_2O emissions through mechanisms such as pH elevation, substrate adsorption, enhanced aeration, and the promotion of N_2O reduction. A meta-analysis encompassing 112 studies demonstrated that biochar reduces N_2O emissions by an average of 54% [83]. Field experiments in China's winter wheat-summer maize rotation system [84] confirmed a 38–45% reduction in annual N_2O emissions over a 3–4 year period with the application of 20–40 t/ha of biochar.

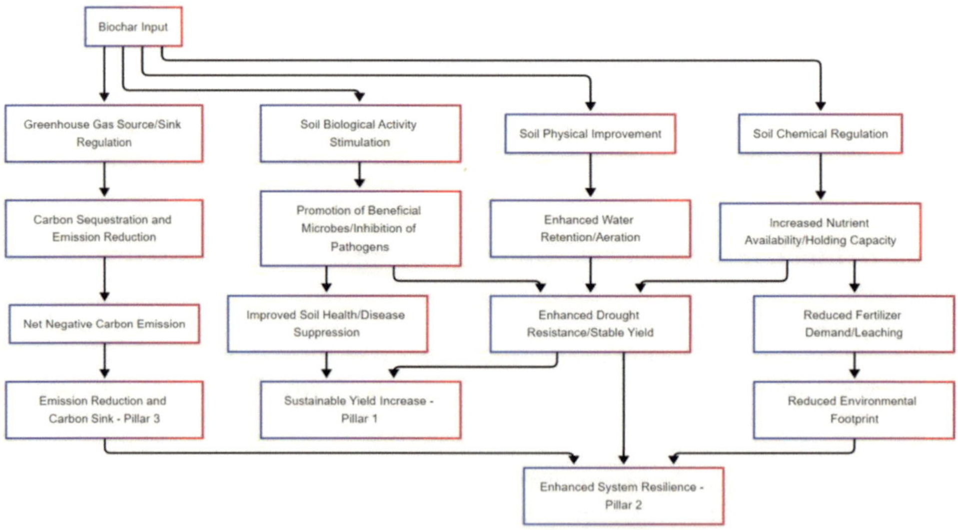

Figure 2.
Multilevel action framework of biochar in empowering climate-smart agriculture (CSA) three pillars.

b. CH_4 flux regulation: The impact of biochar on CH_4 emissions is complex and depends on soil moisture and properties. In paddy fields, low doses of biochar can reduce CH_4 emissions by promoting oxidizing bacteria, while high doses or pore-blocking effects may lead to increased emissions. Meta-analysis [85] indicated an insignificant impact on upland CH_4, with variable effects in paddy fields (ranging from −37% to +66%), but overall, there was a significant reduction in the net greenhouse effect (GWP).

c. Indirect emission reduction: Biochar enhances nutrient use efficiency (NUE), which decreases the demand for chemical nitrogen and the associated production and transport emissions [86, 87]. Improved root growth and photosynthetic efficiency further enhance crop carbon assimilation.

4.3 Enhanced resilience: Fundamental support for drought resistance, yield protection, and nutrient efficiency

4.3.1 Water use efficiency (WUE)

The porous structure of biochar (characterized by high surface area and porosity) improves soil water-holding capacity (WHC), thereby increasing the available water content (AWC), particularly in sandy soils.

a. Improved water retention: Biochar can enhance soil AWC by 10–50%, with high-temperature biochar typically yielding better performance [87].

b. Reduced evapotranspiration and optimized transpiration: Biochar diminishes direct soil evaporation and fosters deep root growth, optimizing transpiration (T/ET ratio) and WUE. In arid and semi-arid regions, the application of biochar (20 t/ha) resulted in a 25% increase in maize WUE and an 18–30% increase in wheat WUE [88].

c. Long-term trials in Colombia [89] demonstrated that biochar improved maize survival and yield stability during the dry season.

4.3.2 Nutrient cycle efficiency and pollution control

a. Enhanced nutrient retention: Biochar's surface functional groups and cation exchange capacity (CEC) effectively adsorb NH_4^+, K^+, Ca^{2+}, and NH_2^+, thereby reducing leaching. Its pores also capture dissolved organic carbon (DOC) and anions (PO_4^{3-}), ensuring nutrient availability in highly leached tropical soils [66].

b. Improved nutrient use efficiency (NUE): By minimizing losses and enhancing the rhizosphere microenvironment, biochar boosts nutrient absorption, leading to higher yields with equal or reduced fertilizer inputs. Meta-analysis indicates a 15–20% increase in NUE and a 10–25% increase in phosphorus use efficiency (PUE) [87, 90].

c. Non-point source pollution control: The adsorption of nutrients and organic pollutants (such as pesticides and antibiotics) mitigates the risk of migration to water bodies [91].

4.3.3 Enhanced soil health and biological activity

Biochar offers physical protection and carbon sources for soil microorganisms, leading to increased microbial biomass, diversity, and activity, including enzyme activity. Robust microbial communities are essential for nutrient cycling, soil structure stability, and disease suppression. Certain biochars can directly inhibit soil-borne pathogens (e.g., Fusarium, Rhizoctonia) and induce plant systemic resistance (ISR), thereby reducing dependence on pesticides [92].

4.4 Intelligent regulation: Biochar as a "sensing-response" carrier in CSA

The value of biochar extends beyond its inherent improvement properties; it also encompasses its integration with modern information technologies (sensors, IoT, AI) as an "intelligent carrier," facilitating dynamic sensing and precision regulation of CSA systems:

a. Integration of biochar-based sensors: Miniature environmental sensors (such as temperature/humidity, pH, and ion-selective electrodes) embedded in biochar particles monitor soil moisture, salinity, nutrients (NO_3^-, K^+), and pollutants in real time [93].

b. IoT and data-driven decision-making: Data collected from biochar-based sensors is aggregated on cloud platforms, where AI models (including machine learning and deep learning) generate variable irrigation and fertilization prescriptions, predict crop stress, and optimize management strategies [17].

c. Closed-loop feedback regulation: AI-driven automatic control systems implement precise water and fertilizer supply, with biochar enhancing soil buffering capacity for stable and effective regulation [17].

4.5 Typical cases: Validation from theory to practice

Dryland Agriculture in China's Loess Plateau: Large-scale field trials [94] demonstrated that straw biochar (15–30 t/ha) combined with film mulching:

a. Increased soil organic carbon by 20–50% and water retention by 15–30%.

b. Reduced spring maize irrigation needs by 1–2 times during critical growth stages, resulting in improved WUE by 20–35%.

c. Maintained relatively high yields during extreme drought years, significantly enhancing system resilience.

Australia's "Soil Carbon Initiative" and Carbon Agriculture Projects: Established agricultural carbon credit (ACCU) market mechanisms enable farmers to engage in biochar projects, verifying carbon sequestration and emission reduction benefits through scientific monitoring, and trading carbon credits for profit [95, 96]. Blockchain technology tracks biochar origin and carbon credit generation, ensuring transparency.

DOI: http://dx.doi.org/10.5772/intechopen.1011847

4.6 Challenges and optimization pathways

Cost-effectiveness: The costs associated with raw material collection, pyrolysis, transportation, and field application remain significant barriers to scaling. There is a need to optimize pyrolysis technology (mobile, small-scale, combined heat and power), develop high-value applications, and leverage revenues from carbon markets.

Effect variability: The effects of biochar are influenced by feedstock, pyrolysis conditions, soil type, climate, and crop management. The development of big data and AI-based predictive models is essential for "site-specific" and "material-specific" precision solutions [97].

Long-term effect monitoring: There is a need for more long-term (>10 years) agronomic and environmental effect data under various management practices.

Standardization and Policy Integration: The lack of unified biochar product quality standards, carbon accounting methods, and certification systems poses challenges. Urgent integration of biochar into national and regional CSA strategies and agricultural subsidy systems is necessary.

4.7 Conclusion: Toward the core of smart agricultural systems

Biochar has evolved from a mere soil amendment to a fundamental component of Climate-Smart Agriculture (CSA) systems. By facilitating irreversible carbon sequestration, significantly reducing greenhouse gas (GHG) emissions (particularly N_2O), revolutionizing water and nutrient efficiency management, and comprehensively improving soil health and biological activity, it provides robust scientific and technological support for the three pillars of CSA. Notably, biochar's integration with modern information technologies (IoT, sensor networks, AI) as a functional intelligent carrier ushers in a new paradigm of precision sensing, data-driven decision-making, and dynamic feedback regulation in smart agriculture. Global cases, such as efficient dryland practices in China's Loess Plateau and innovations in Australia's carbon agriculture market, underscore its technical feasibility and numerous benefits. Despite challenges related to cost optimization, effect prediction, long-term monitoring, and policy standard integration, biochar—with its unique cross-scale functions and strong system integration potential—has become an indispensable engineering tool for advancing agriculture toward a future characterized by high yields, efficiency, resource conservation, environmental sustainability, and resilience. Its comprehensive integrated application signifies a paradigm shift from passive response to active design and regulation of sustainable eco-production systems in agriculture.

5. Future outlook: Artificial intelligence and policy innovation—Toward an intelligent, fair, and resilient biochar system

Biochar's role in smart agriculture and environmental regulation is evolving, presenting a new phase filled with opportunities and challenges. Its transition from "passive amendment" to "active intelligent regulatory core" relies heavily on the profound integration of artificial intelligence (AI) and the collaborative innovation of policy systems. This section delineates key future pathways, concentrating on AI-driven technological advancements and the collaborative design of policy, ethics, and market frameworks to maximize biochar's potential in tackling global environmental challenges and supporting sustainable agriculture.

5.1 AI-driven biochar systems: From precision to prediction, from response to adaptation

Future applications of biochar will be intricately woven into the "nervous system" of the "Smart Earth," with AI functioning as its "brain" to achieve unprecedented levels of precision, predictability, and adaptability.

5.1.1 AI-optimized biochar "customized" design and production

Multi-Objective Joint Optimization Model: Moving beyond single-performance optimization (e.g., carbon sequestration, adsorption capacity), we will develop AI models leveraging reinforcement learning and multi-objective evolutionary algorithms. These models will consider target application scenarios (e.g., enhancing water retention in sandy soils, reducing cadmium in acidic soils, and enriching greenhouse CO_2), available raw material properties, cost constraints, and environmental regulatory requirements. The model will automatically identify optimal pyrolysis process parameters (temperature curve, heating rate, residence time, catalyst utilization) and potential modification strategies (e.g., nano-iron loading, microbial inoculation), generating biochar design schemes that meet specific "performance envelopes" (e.g., high stability, high phosphorus adsorption, low PAHs generation) [98].

Real-Time Process Control and Quality Prediction: The deployment of multi-modal sensors (infrared thermal imaging, gas composition mass spectrometry, pressure vibration sensing) in pyrolysis reactors will facilitate the use of deep learning (CNN, LSTM) to analyze real-time data streams. This approach will accurately predict reaction progress, key product characteristics (specific surface area, functional group distribution), and dynamics of potential pollutant (PAHs) generation, providing real-time feedback to regulate heating power, feed rate, and carrier gas flow, thereby achieving "self-perception-self-decision-self-optimization" in intelligent production [99].

5.1.2 Constructing biochar-environment-crop "digital twins"

Multi-scale, multi-physics coupled simulation models will be developed, encompassing microscale (reactions within biochar pores, interactions between microorganisms and char surfaces), mesoscale (carbon-water-nutrient flow in rhizosphere microdomains), and macroscale (water-carbon fluxes at field and regional scales). These models will integrate physical (water flow, heat conduction), chemical (adsorption-desorption, redox), and biological (microbial community dynamics, crop physiology) processes. High-performance computing and artificial intelligence (e.g., physics-informed neural networks, PINNs) will be utilized to tackle complex coupling challenges, facilitating ultra-high-precision simulations of system behaviors following biochar application [100].

Farmers or managers will be able to input localized data related to soil, climate, crops, management practices, and proposed biochar parameters into the digital twin platform. Utilizing twin simulations, the platform will forecast the long-term impacts (5–10-year scale) of biochar application on crop yield, WUE, nitrogen and phosphorus loss risk, net greenhouse gas balance, and soil health indicators across various scenarios (e.g., normal years, droughts, heavy rainfall), providing visual comparisons and optimal decision-making recommendations.

5.1.3 AI-enabled real-time closed-loop control systems

A dense sensing network will be established across agricultural fields through the deployment of high-density, low-cost, low-power multi-parameter soil sensors (measuring moisture, temperature, pH, electrical conductivity, and specific ions such as NO_3- and NH_4^+), crop physiological sensors (monitoring stem flow, leaf temperature, and canopy spectroscopy), and environmental sensors (including weather stations and groundwater monitoring). Lightweight AI models (e.g., TinyML) will be implemented at field gateways or edge computing nodes to process sensor data streams in real time. Based on predefined optimization goals (e.g., maximizing water use efficiency, minimizing nitrogen leaching), along with predictions from digital twins, the models will dynamically generate control instructions: automatically adjusting the switch and flow rate of drip irrigation valves (facilitating biochar-water synergistic regulation), controlling intelligent fertilizer dispensers to deliver liquid fertilizers on demand (enabling biochar-nutrient synergistic regulation), and regulating ventilation in greenhouses to optimize CO_2 concentration (supporting biochar mineralization for CO_2 supply). This establishes a second/minute-level closed-loop system of "perception-analysis-decision-execution," maximizing the spatial and temporal efficiency of biochar regulation [101].

5.2 Policy innovation: Developing an incentive-compatible, equitable, and risk-controlled governance ecosystem

The realization of technological potential is intrinsically linked to proactive and systematic policy design and institutional safeguards. Future policies must achieve groundbreaking innovations across four dimensions: incentives, equity, risk, and collaboration.

5.2.1 Smart policy toolkit

A high-credibility monitoring, reporting, and verification (MRV) platform for biochar carbon sequestration and emission reduction will be established using remote sensing (satellite/UAV monitoring of crop growth and vegetation coverage changes), Internet of Things (IoT) technologies (real-time monitoring of soil carbon pools and N_2O emissions), and blockchain (providing tamper-proof records of biochar production, transportation, and application data). AI models will integrate multi-source data to automatically and accurately account for the actual carbon sequestration and emission reduction benefits at the plot scale, enabling a shift from "input-based" to "result-based" subsidy distribution and carbon credit issuance, thereby significantly enhancing policy efficiency and fairness (**Figure 3**) [102].

A "policy simulator" based on Agent-Based Modeling (ABM) and system dynamics will be developed to simulate the behavioral responses of stakeholders (e.g., farmers, enterprises, consumers) under various policy combinations (e.g., carbon tax rates, subsidy intensities, biochar standard requirements). This simulator will predict the short-term and long-term impacts of policies on biochar market penetration, environmental benefits (such as carbon sequestration and water quality), social equity (including smallholder participation), and industrial development, providing a "sandbox simulation" for policy formulation and reducing trial-and-error costs [103].

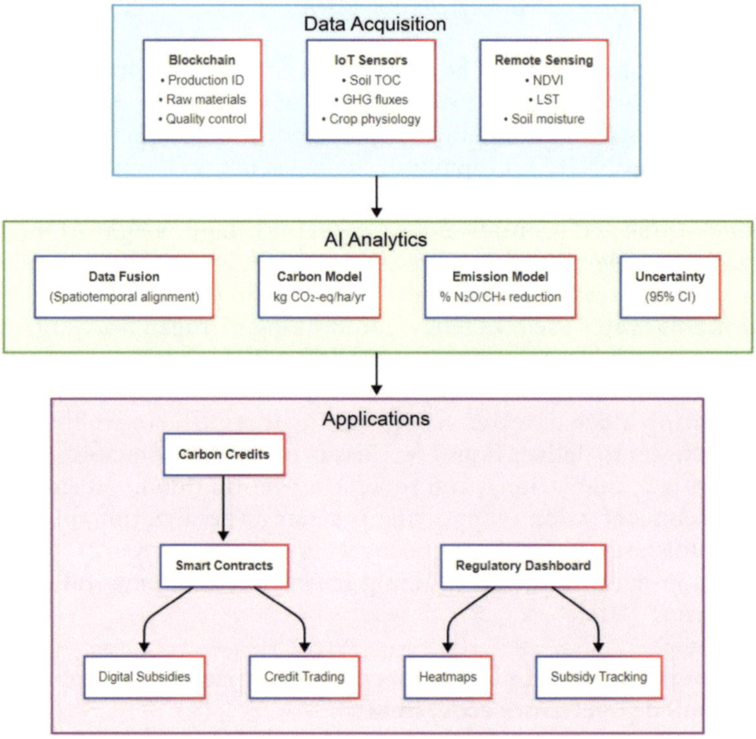

Figure 3.
Precision carbon sequestration MRV system for biochar based on AI and blockchain.

5.2.2 Embedding equity and inclusive mechanisms

A "Universal Biochar" initiative will be launched, establishing dedicated funds to support the research, development, and promotion of modular, low-cost, low-emission small/mini pyrolysis equipment designed for smallholders (e.g., solar-assisted heating). Community sharing models, such as biochar production cooperatives, will be developed to minimize investment costs for individual farmers. Policies will be implemented to provide smallholders with technical training, initial equipment subsidies, or microcredit guarantees, ensuring they can easily access carbon benefits through the intelligent monitoring, reporting, and verification (MRV) system outlined above [104].

In international carbon credit transactions and large-scale biochar investment projects, mandatory fair trade certification or community benefit-sharing funds will be required. The rights and interests of local communities, particularly indigenous peoples and smallholders, regarding land rights, resource rights, free, prior, and informed consent (FPIC), and carbon income sharing (e.g., a minimum of 20%) will be clearly stipulated, with independent third-party supervision mechanisms established [105].

Governments and non-governmental organizations will collaborate to provide digital skills and biochar knowledge training for farmers, especially in resource-constrained areas. Multilingual, user-friendly AI-assisted decision-making mobile applications will be developed to offer farmers free, locally optimized biochar application advice and risk alerts.

5.2.3 Dynamic risk governance and adaptive regulation

For the application of biochar in conjunction with new technologies (e.g., nano-modification, genetic engineering bacteria coupling), an "adaptive regulatory sandbox" facilitated by AI will be established. Innovative trials will be permitted within a controlled scope and timeframe. AI-driven real-time monitoring networks (see Section 6) will comprehensively collect environmental, health, and social impact data. Regulatory agencies will dynamically adjust regulatory requirements—such as modifying application limits and adding monitoring indicators—based on real-time data and AI risk prediction models, achieving a "regulate while developing, learn while regulating" approach [106].

The insurance industry will be encouraged to develop AI-based risk assessment models for long-term (e.g., 30-year) environmental liability insurance products for biochar. Premiums may fluctuate dynamically according to plot risk levels (soil properties, groundwater vulnerability, biochar pollutant levels) and enterprise risk management levels (e.g., the adoption of intelligent monitoring). Blockchain technology will ensure the traceability of historical data and liability determination.

An international organization, such as the FAO or UNEP, will lead the establishment of an open global biochar application case database, with a particular emphasis on long-term monitoring data (over 10 years) and unforeseen negative events. AI will be employed for big data mining to identify potential emerging risk signals, thereby establishing cross-border risk information sharing and early warning mechanisms.

5.2.4 Promoting cross-sectoral collaboration and global governance

A domestic "Biochar-Climate-Agriculture-Environment" policy coordination office will be established to dismantle departmental silos across agriculture, environmental protection, climate, science and technology, and industry. This office will also create a high-level coordination agency. Its primary objective will be to comprehensively formulate national biochar strategies, ensuring that agricultural subsidy policies, climate change commitments (NDCs), soil protection regulations, circular economy goals, and biochar industry development policies are seamlessly integrated and aligned with mutual incentives.

A Global Biochar Innovation and Governance Alliance (GBIGA) will be initiated, establishing a global coalition of diverse stakeholders, including governments, research institutions, enterprises, NGOs, and farmer organizations. Its core responsibilities will encompass: (1) promoting open science and sharing research data and models; (2) coordinating international standard mutual recognition (e.g., EBC, IBI); (3) establishing fair and transparent global carbon accounting methodologies; (4) facilitating technology transfer to developing countries and capacity building; (5) collaboratively addressing cross-border environmental risks (such as biochar particle migration); and (6) negotiating governance rules to prevent carbon leakage and carbon colonialism [107].

Biochar applications that adhere to high standards—such as sustainable feedstocks, low-carbon processes, strict pollution control, and equity—will be actively promoted for formal inclusion in the Nationally Determined Contributions (NDCs) under the United Nations Framework Convention on Climate Change (UNFCCC), the Carbon Offsetting and Reduction Scheme for International

Aviation (CORSIA), the EU Carbon Border Adjustment Mechanism (CBAM), and other global climate policy tools as qualified methods for emission reduction and removal. This will create a stable and expansive international demand market for biochar [108].

6. Conclusions: Co-shaping a smart and sustainable future from soil to system

Biochar, a material rooted in ancient charcoal-burning practices, is experiencing a significant transformation driven by disruptive technologies such as artificial intelligence, as well as innovative policies. It is evolving from a simple soil amendment to a central component of smart agriculture and environmental sustainability regulation systems. This chapter systematically outlines the pathway, mechanisms, and extensive prospects of this transformation while thoroughly analyzing the associated challenges.

a. Core essence of the paradigm shift: The value of biochar transcends its inherent physical and chemical properties (such as porosity, alkalinity, and adsorption) and now encompasses its new roles as an intelligent carrier (integrating sensors and nanomaterials), a system connector (linking carbon, water, and nutrient cycles), and an information node (blockchain recording lifecycle data). Biochar serves as a crucial bridge between the physical world (soil, crops, environment) and the digital landscape (AI models, IoT, blockchain).

b. Transformative power of triple synergy:

1. Technological synergy (AI + Biochar): AI is deeply integrated throughout the biochar value chain, including design, production, monitoring, simulation, decision-making, and closed-loop control. This integration represents a shift from "experience-driven" to "data and model-driven" processes, unlocking unprecedented levels of precision, predictability, and adaptability.

2. System synergy (carbon-water-nutrient closed loop): As a central medium, biochar reconstructs the "carbon-water-nutrient" cycle within the smart agriculture framework, enabling multi-objective collaborative optimization of resource efficiency (water and fertilizer conservation), environmental emission reduction (carbon sequestration, N_2O reduction, pollution control), and productivity enhancement. This robustly supports the implementation of Climate-Smart Agriculture (CSA).

3. Governance synergy (policy-market-ethics): Blockchain technology ensures transparent traceability and trust, while innovative policy tools (intelligent subsidies, carbon MRV, adaptive regulation) create effective incentives and risk prevention measures. Integrating fairness, ethics, and social participation into standards and governance frameworks is essential for ensuring the equitable distribution of technological benefits and preventing new forms of inequality. The convergence of the "circular economy and blockchain" fosters the resource utilization of biomass waste and the green transformation of industries.

d. Call to action for the future: To effectively transform biochar from potential into reality, it is imperative to:

1. Continuously enhance fundamental research and technological innovation, focusing on long-term behaviors under complex environmental interactions, assessing environmental health risks associated with new biochars (e.g., nano-biochars), and developing more efficient and inclusive AI models and hardware.

2. Accelerate the iterative cycle of policy, market, and technology by rapidly refining incentive-compatible, equitable, effective, and risk-controllable policy frameworks and market mechanisms through "policy laboratories" and regulatory sandbox pilots, thereby facilitating large-scale applications.

3. Commit to responsible innovation by prioritizing ethical reviews, social impact assessments (SIA), fair benefit sharing (FPIC), smallholder empowerment, and global justice in technology development and project implementation, fostering broad public trust.

4. Strengthen global collaboration and knowledge sharing by establishing a robust global alliance (such as GBIGA), collaboratively formulating regulations, sharing knowledge, coordinating actions, and bridging gaps to ensure biochar technology serves as a global public good that benefits all nations, particularly the most vulnerable regions and communities.

The synergy of biochar with smart agriculture and environmental practices presents a pathway toward a more resilient, efficient, and equitable agricultural and ecological system. This journey is fraught with scientific uncertainties, technical complexities, competing interests, and ethical considerations. Nevertheless, by deeply integrating advanced intelligent technologies, innovative policy frameworks, and a steadfast commitment to sustainable development and equity, humanity can transform biochar—an ancient material—into a pivotal tool for cultivating a smart, green, and inclusive future. From the microcosm of soil to the global ecosystem, a profound, sustainable revolution driven by biochar is unfolding, illuminated by the light of wisdom.

Acknowledgements

This chapter was supported by Jinan City-University Integration Development Strategy Project (JNSX2024025), Excellent Doctoral Program in University of Jinan (XBS2440 and XBS2439).

Conflict of interest

The authors declare no conflict of interest.

Author details

Mengmeng Kong[1†], Xiaoyong Liu[1*†], Dongjie Chen[2], Yangyang Xu[1], Jie Zhang[1], Xiting Chen[1], Jinyuan Liang[3] and Hui-lian Xu[1*]

1 School of Biological Science and Technology, University of Jinan, Jinan, China

2 National Engineering Research Center for Agricultural Products Logistics, Shandong Institute of Commerce and Technology, Jinan, China

3 School of Chemistry and Chemical Engineering, University of Jinan, Jinan, China

*Address all correspondence to: bio_liuxy@ujn.edu.cn and bio_xuhl@ujn.edu.cn

† These authors contribute equally.

References

[1] Glaser B, Haumaier L, Guggenberger G, et al. The 'Terra Preta' phenomenon: A model for sustainable agriculture in the humid tropics. Die Naturwissenschaften. 2001;**88**(1):37-41. DOI: 10.1007/s001140000193

[2] Lehmann J, Joseph S, editors. Biochar for Environmental Management: Science, Technology and Implementation. London: Routledge; 2015

[3] Wang D, Fonte SJ, Parikh SJ, et al. Biochar additions can enhance soil structure and the physical stabilization of C in aggregates. Geoderma. 2017;**303**:110-117. DOI: 10.1016/j.geoderma.2017.05.027

[4] Amir R, Jahan S, Zafar M, et al. Organic amendments optimize *Triticum aestivum* var Zincol-16 performance under water stress through enhancement in physio biochemical traits, secondary metabolites and yield. Scientific Reports. 2025;**15**:17995. DOI: 10.1038/s41598-025-99506-7

[5] Laird DA, Novak JM, Collins HP, et al. Multi-year and multi-location soil quality and crop biomass yield responses to hardwood fast pyrolysis biochar. Geoderma. 2017;**289**:46-53. DOI: 10.1016/j.geoderma.2016.11.025

[6] Olmo M, Rafael V, Salazar P, et al. Changes in soil nutrient availability explain biochar's impact on wheat root development. Plant and Soil. 2016;**399**:333-343. DOI: 10.1007/s11104-015-2700-5

[7] Xia H, Riaz M, Zhang M, et al. Biochar increases nitrogen use efficiency of maize by relieving aluminum toxicity and improving soil quality in acidic soil. Ecotoxicology and Environmental Safety. 2020;**196**:110531. DOI: 10.1016/j.ecoenv.2020.110531

[8] Schmidt HP, Kammann C, Hagemann N, et al. Biochar in agriculture-A systematic review of 26 global meta-analyses. GCB Bioenergy. 2021;**13**(11):1708-1730. DOI: 10.1111/gcbb.12889

[9] Wang JY, Xiong ZQ, Kuzyakov Y. Biochar stability in soil: Meta-analysis of decomposition and priming effects. Global Change Biology Bioenergy. 2016;**8**(3):512-523. DOI: 10.1111/gcbb.12266

[10] Glaser B, Birk JJ. State of the scientific knowledge on properties and genesis of Anthropogenic Dark Earths in Central Amazonia (terra preta de Índio). Geochimica et Cosmochimica Acta. 2012;**82**:39-51. DOI: 10.1016/j.gca.2010.11.029

[11] International Biochar Initiative. Biochar Standards. 2015. Available from: https://www.biochar-international.pdf [Accessed: June 9, 2025]

[12] IPCC. Climate Change and Land: An IPCC Special Report on Climate Change, Desertification, Land Degradation, Sustainable Land Management, Food Security, and Greenhouse Gas Fluxes in Terrestrial Ecosystems. 2019. Available from: https://www.ipcc.ch/srccl.pdf [Accessed: June 9, 2025]

[13] Tan Z, Lin CSK, Ji X, et al. Returning biochar to fields: A review. Applied Soil Ecology. 2017;**116**:1-11. DOI: 10.1016/j.apsoil.2017.03.017

[14] FAO. The Future of Food and Agriculture-Trends and Challenges. Rome: The Food and Agriculture Organization of the United Nations

(FAO); 2017. [Internet]. Available from: https://www.fao.org/3/i6583e/i6583e.pdf [Accessed: June 9, 2025]

[15] Costa G, Forestiero A, Gentile AF, et al. An AI-driven architecture for precision agriculture: IoT, machine learning, and digital twin integration for sustainable crop protection. In: International Conference on Big Data (BigData). USA: IEEE; 2024. pp. 4685-4469

[16] Zhang D, Yan M, Niu Y, et al. Is current biochar research addressing global soil constraints for sustainable agriculture? Agriculture, Ecosystems & Environment. 2016;**226**:25-32. DOI: 10.1016/j.agee.2016.04.010

[17] Joseph S, Cowie AL, Van Zwieten L, et al. How biochar works, and when it doesn't: A review of mechanisms controlling soil and plant responses to biochar. GCB Bioenergy. 2021;**13**(11):1731-1764. DOI: 10.1111/gcbb.12885

[18] Huang C, Chen Y, Jin L, et al. Properties of biochars derived from different straw at 500°C pyrolytic temperature: Implications for their use to improving acidic soil water retention. Agricultural Water Management. 2024;**301**:108953. DOI: 10.1016/j.agwat.2024.108953

[19] Kim SA, Kim EB, Imran M, et al. Naturally manufactured biochar materials based sensor electrode for the electrochemical detection of polystyrene microplastics. Chemosphere. 2024;**51**:141151. DOI: 10.1016/j.chemosphere.2024.141151

[20] Liu X, Antonietti M. Molten salt activation for synthesis of porous carbon nanostructures and carbon sheets. Carbon. 2014;**69**:460-466. DOI: 10.1016/j.carbon.2013.12.049

[21] Li B, Li M, Xie X, et al. Pyrolysis of rice husk in molten lithium chloride: Biochar structure evolution and CO_2 adsorption. Journal of the Energy Institute. 2024;**113**:101526. DOI: 10.1016/j.joei.2024.101526

[22] Adiraju A, Brahem A, Lu T, et al. Electrochemical enrichment of biocharcoal modified on carbon electrodes for the detection of nitrite and paraxon ethyl pesticide. Journal of Composites Science. 2024;**8**(6):217. DOI: 10.3390/jcs8060217

[23] Sun Z, Wei H, Guo F, et al. Study on the characteristics of microwave pyrolysis of pine wood catalyzed by bimetallic catalysts prepared with microwave-assisted hydrothermal char. Journal of Analytical and Applied Pyrolysis. 2024;**181**:106597. DOI: 10.1016/j.jaap.2024.106597

[24] Yuan P, Wang J, Pan Y, et al. Review of biochar for the management of contaminated soil: Preparation, application and prospect. Science of the Total Environment. 2019;**659**:473-490. DOI: 10.1016/j.scitotenv.2018.12.400

[25] Xiang W, Zhang X, Chen J, et al. Biochar technology in wastewater treatment: A critical review. Chemosphere. 2020;**252**:126539. DOI: 10.1016/j.chemosphere.2020.126539

[26] Mukherjee A, Lal R, Zimmerman AR. Effects of biochar and other amendments on the physical properties and greenhouse gas emissions of an artificially degraded soil. Science of the Total Environment. 2014;**487**:26-36. DOI: 10.1016/j.scitotenv.2014.03.141

[27] Australian Government, Clean Energy Regulator. Emissions Reduction Fund: Soil Carbon Projects. 2023. Available from: https://www.cleanenergyregulator.gov.au/ERF/

Choosing-a-project-type/Opportunities-for-the-land-sector/Soil-Carbon.pdf [Accessed: June 9, 2025]

[28] Zhang P, Zhang T, Zhang J, et al. A machine learning assisted prediction of potential biochar and its applications in anaerobic digestion for valuable chemicals and energy recovery from organic waste. Carbon Neutrality. 2024;**3**(1):2. DOI: 10.1007/s43979-023-00078-0

[29] Woolf D, Amonette JE, Street-Perrott FA, et al. Sustainable biochar to mitigate global climate change. Nature Communications. 2010;**1**(1):56. DOI: 10.1038/ncomms1053

[30] Mishra H, Mishra D. AI for data-driven decision-making in smart agriculture: From field to farm management. In: Artificial Intelligence Techniques in Smart Agriculture. Singapore: Springer Nature Singapore; 2024. pp. 173-193. DOI: 10.1007/978-981-97-5878-4_11

[31] Nighojkar A, Pandey S, Naebe M, et al. Using machine learning to predict the efficiency of biochar in pesticide remediation. npj Sustainable Agriculture. 2023;**1**(1):1. DOI: 10.1038/s44264-023-00001-1

[32] Zhang Q, Song Y, Wu Z, et al. Effects of six-year biochar amendment on soil aggregation, crop growth, and nitrogen and phosphorus use efficiencies in a rice-wheat rotation. Journal of Cleaner Production. 2020;**242**:118435. DOI: 10.1016/j.jclepro.2019.118435

[33] Lehmann J, Rillig MC, Thies J, et al. Biochar effects on soil biota: A review. Soil Biology and Biochemistry. 2011;**43**:1812-1836. DOI: 10.1016/j.soilbio.2011.04.022

[34] Xiao X, Chen B, Zhu L. Transformation, morphology, and dissolution of silicon and carbon in rice straw-derived biochars under different pyrolytic temperatures. Environmental Science & Technology. 2014;**48**(6):3411-3419. DOI: 10.1021/es405676h

[35] Kimetu JM, Lehmann J. Stability and stabilisation of biochar and green manure in soil with different organic carbon contents. Soil Research. 2010;**48**(7):577-585. DOI: 10.1071/SR10036.1

[36] Chen Z, Liu T, Dong JF, et al. Sustainable application for agriculture using biochar-based slow-release fertilizers: A review. ACS Sustainable Chemistry & Engineering. 2022;**11**(1):1-12. DOI: 10.1021/acssuschemeng.2c05691

[37] Rajput VD, Minkina T, Ahmed B, et al. Nano-biochar: A novel solution for sustainable agriculture and environmental remediation. Environmental Research. 2022;**210**:112891. DOI: 10.1016/j.envres.2022.112891

[38] Rajapaksha AU, Chen SS, Tsang DCW, et al. Engineered/designer biochar for contaminant removal/immobilization from soil and water: Potential and implication of biochar modification. Chemosphere. 2016;**148**:276-291. DOI: 10.1016/j.chemosphere.2016.01.043

[39] Wang B, Gao B, Fang J. Recent advances in engineered biochar productions and applications. Critical Reviews in Environmental Science and Technology. 2017;**47**(22):2158-2207. DOI: 10.1080/10643389.2017.1418580

[40] Li R, Wang JJ, Zhou B, et al. Recovery of phosphate from aqueous solution by magnesium oxide decorated magnetic biochar and its potential as phosphate-based fertilizer substitute. Bioresource Technology. 2016;**215**:209-214. DOI: 10.1016/j.biortech.2016.02.125

[41] Zhang M, Gao B, Yao Y, et al. Synthesis of porous MgO-biochar nanocomposites for removal of phosphate and nitrate from aqueous solutions. Chemical Engineering Journal. 2012;**210**:26-32. DOI: 10.1016/j.cej.2012.08.052

[42] Yao Y, Gao B, Wu F, et al. Engineered biochar from biofuel residue: Characterization and its silver removal potential. ACS Applied Materials & Interfaces. 2015;**7**(19):10634-10640. DOI: 10.1021/acsami.5b03131

[43] Su H, Fang Z, Tsang PE, et al. Stabilisation of nanoscale zero-valent iron with biochar for enhanced transport and in-situ remediation of hexavalent chromium in soil. Environmental Pollution. 2016;**214**:94-100. DOI: 10.1016/j.envpol.2016.03.072

[44] Nan N, DeVallance DB, Xie X, et al. The effect of bio-carbon addition on the electrical, mechanical, and thermal properties of polyvinyl alcohol/biochar composites. Journal of Composite Materials. 2016;**50**(9):1161-1168. DOI: 10.1177/0021998315589770

[45] Mertens J, Germer J, de Araújo Filho JC, et al. Effect of biochar, clay substrate and manure application on water availability and tree-seedling performance in a sandy soil. Archives of Agronomy and Soil Science. 2017;**63**(7):969-983. DOI: 10.1080/03650340.2016.1249473

[46] Chen S, Yang M, Ba C, et al. Preparation and characterization of slow-release fertilizer encapsulated by biochar-based waterborne copolymers. Science of the Total Environment. 2018;**615**:431-437. DOI: 10.1016/j.scitotenv.2017.09.209

[47] Li Y, Xu R, Wang H, et al. Recent advances of biochar-based electrochemical sensors and biosensors. Biosensors. 2022;**12**(6):377. DOI: 10.3390/bios12060377.12

[48] Pudza MY, Abidin ZZ, Rashid SA, et al. Eco-friendly sustainable fluorescent carbon dots for the adsorption of heavy metal ions in aqueous environment. Nanomaterials. 2020;**10**(2):315. DOI: 10.3390/nano10020315

[49] Yue N, Wu J, Qi W, et al. Algae-derived biochar nanozyme array for discrimination and detection of multiple pesticides in soil, water and food. Food Chemistry. 2024;**438**(000):9. DOI: 10.1016/j.foodchem.2023.137946

[50] Liu K, Fan D, Zou S, et al. Capacitance capacity of biochar-incorporated cementitious composite. Resources, Conservation and Recycling. 2025;**218**:108223. DOI: 10.1016/j.resconrec.2025.108223

[51] El-Naggar A, Lee SS, Rinklebe J, et al. Biochar application to low fertility soils: A review of current status, and future prospects. Geoderma. 2019;**337**:536-554. DOI: 10.1016/j.geoderma.2018.09.034

[52] Singh H, Northup BK, Rice CW, et al. Biochar applications influence soil physical and chemical properties, microbial diversity, and crop productivity: A meta-analysis. Biochar. 2022;**4**(1):8. DOI: 10.1007/s42773-022-00138-1

[53] Kookana RS, Sarmah AK, Zwieten LV, et al. Biochar application to soil: Agronomic and environmental benefits and unintended consequences. Advances in Agronomy. 2011;**112**:103-143. DOI: 10.1016/B978-0-12-385538-1.00003-2

[54] Baykal G, Kazan D, Seval K, et al. Determination of lead and cadmium in water samples by magnetic

solid-phase extraction with iron oxide@ silicon oxide-graphene oxide (Fe_3O_4@ SiO_2-GO) hybrid magnetic nanoparticles and microinjection sampling flame atomic absorption spectrometry. Instrumentation Science & Technology. 2021;**50**(3):288-305. DOI: 10.1080/10739149.2021.1996389

[55] Samoraj M, Mironiuk M, Witek-Krowiak A, et al. Biochar in environmental friendly fertilizers-prospects of development products and technologies. Chemosphere. 2022;**296**:133975. DOI: 10.1016/j.chemosphere.2022.133975

[56] Tung TT, Yoo J, Alotaibi FK, et al. Graphene oxide-assisted liquid phase exfoliation of graphite into graphene for highly conductive film and electromechanical sensors. ACS Applied Materials & Interfaces. 2016;**8**(25):16521-16532. DOI: 10.1021/acsami.6b04872

[57] Lehmann J, Gaunt J, Rondon M. Bio-char sequestration in terrestrial ecosystems-A review. Mitigation and Adaptation Strategies for Global Change. 2006;**11**:403-427. DOI: 10.1007/S11027-005-9006-5

[58] Zimmerman AR. Abiotic and microbial oxidation of laboratory-produced black carbon (biochar). Environmental Science & Technology. 2010;**44**(4):1295-1301. DOI: 10.1021/es903140c

[59] Cayuela ML, Van Zwieten L, Singh BP, et al. Biochar's role in mitigating soil nitrous oxide emissions: A review and meta-analysis. Agriculture Ecosystems & Environment. 2014;**191**:5-16. DOI: 10.1016/j.agee.2013.10.009

[60] Borchard N, Schirrmann M, Luz Cayuela M, et al. Biochar, soil and land-use interactions that reduce nitrate leaching and N_2O emissions: A meta-analysis. Science of the Total Environment. 2018;**651**(PT.2):2354-2364. DOI: 10.1016/j.scitotenv.2018.10.060.7

[61] Jeffery S, Verheijen FGA, Kammann C, et al. Biochar effects on methane emissions from soils: A meta-analysis. Soil Biology & Biochemistry. 2016;**101**:251-258. DOI: 10.1016/j.soilbio.2016.07.021

[62] Herath HMSK, Camps-Arbestain M, Hedley M. Effect of biochar on soil physical properties in two contrasting soils: An Alfisol and an Andisol. Geoderma. 2013;**209**:188-197. DOI: 10.1016/j.geoderma.2013.06.016.9

[63] Omondi MO, Xia X, Nahayo A, et al. Quantification of biochar effects on soil hydrological properties using meta-analysis of literature data. Geoderma: An International Journal of Soil Science. 2016;**274**:28-34. DOI: 10.1016/j.geoderma.2016.03.029

[64] Yu H, Zou W, Chen J, et al. Biochar amendment improves crop production in problem soils: A review. Journal of Environmental Management. 2019;**232**:8-21. DOI: 10.1016/j.jenvman.2018.10.117

[65] Akhtar SS, Andersen MN, Liu F. Biochar mitigates salinity stress in potato. Journal of Agronomy and Crop Science. 2015;**201**(5):368-378. DOI: 10.1111/jac.12132

[66] Glaser B, Lehmann J, Zech W. Ameliorating physical and chemical properties of highly weathered soils in the tropics with charcoal-a review. Biology and Fertility of Soils. 2002;**35**(4):219-230. DOI: 10.1007/s00374-002-0466-4

[67] Yao Y, Gao B, Zhang M, et al. Effect of biochar amendment on sorption and

leaching of nitrate, ammonium, and phosphate in a sandy soil. Chemosphere. 2012;**89**(11):1467-1471. DOI: 10.1016/j.chemosphere.2012.06.002

[68] Clough T, Condron L, Kammann C, et al. A review of biochar and soil nitrogen dynamics. Agronomy. 2013;**3**(2):275-293. DOI: 10.3390/agronomy3020275

[69] Güereña D, Lehmann J, Hanley K, et al. Nitrogen dynamics following field application of biochar in a temperate North American maize-based production system. Plant & Soil. 2013;**365**(1-2):239-254. DOI: 10.1007/s11104-012-1383-4

[70] DeLuca TH, Gundale MJ, MacKenzie MD, et al. Biochar effects on soil nutrient transformations. In: Biochar for Environmental Management. London: Routledge; 2024:401-440. DOI: 10.4324/9781003297673-16

[71] Barré P, Angers DA, Basile-Doelsch I, et al. Ideas and perspectives: Can we use the soil carbon saturation deficit to quantitatively assess the soil carbon storage potential, or should we explore other strategies? Biogeosciences Discussions. 2017;**2017**:1-12. DOI: 10.5194/bg-2017-395

[72] Ariesen-Verschuur N, Verdouw C, Tekinerdogan B. Digital twins in greenhouse horticulture: A review. Computers and Electronics in Agriculture. 2022;**199**:107183. DOI: 10.1016/j.compag.2022.107183

[73] Sharma D, Tomar GS. Soil nutrient analysis and automatic monitoring in precision agriculture. Communications in Soil Science & Plant Analysis. 2025;**56**(2):145-166. DOI: 10.1080/00103624.2024.2406478

[74] Mohan D, Sarswat A, Ok YS, et al. Organic and inorganic contaminants removal from water with biochar, a renewable, low cost and sustainable adsorbent-a critical review. Bioresource Technology. 2014;**160**(5):191-202. DOI: 10.1016/j.biortech.2014.01.120

[75] Luo S, Wang S, Tian L, et al. Long-term biochar application influences soil microbial community and its potential roles in semiarid farmland. Applied Soil Ecology. 2017;**117-118**:10-15. DOI: 10.1016/j.apsoil.2017.04.024

[76] Wu F, Jia Z, Wang S, et al. Contrasting effects of wheat straw and its biochar on greenhouse gas emissions and enzyme activities in a Chernozemic soil. Biology & Fertility of Soils. 2013;**49**:555-565. DOI: 10.1007/s00374-012-0745-7

[77] Gul S, Whalen JK. Biochemical cycling of nitrogen and phosphorus in biochar-amended soils. Soil Biology and Biochemistry. 2016;**103**(000):15. DOI: 10.1016/j.soilbio.2016.08.001

[78] Zhang A, Liu Y, Pan G, et al. Effect of biochar amendment on maize yield and greenhouse gas emissions from a soil organic carbon poor calcareous loamy soil from Central China Plain. Plant and Soil. 2012;**351**:263-275. DOI: 10.1007/s11104-011-0957-x

[79] Graber ER, Meller Harel Y, Kolton M, et al. Biochar impact on development and productivity of pepper and tomato grown in fertigated soilless media. Plant & Soil. 2010;**337**:481-496. DOI: 10.1007/s11104-010-0544-6

[80] Kaushal A, Yadav RK, Singh N. Chapter 13-Biochar application in sustainable production of horticultural crops in the new era of soilless cultivation. In: Biochar Production for Green Economy. Pittsburgh, USA: Academic Press; 2024. pp. 249-267. DOI: 10.1016/B978-0-443-15506-2.00006-7

[81] Verdouw CN, Robbemond RM, Wolfert J. ERP in agriculture: Lessons learned from the Dutch horticulture. Computers and Electronics in Agriculture. 2015;**114**:125-133. DOI: 10.1016/j.compag.2015.04.002

[82] Lehmann J, Hansel CM, Kaiser C, et al. Persistence of soil organic carbon caused by functional complexity. Nature Geoscience. 2021;**14**(8):625-632. DOI: 10.1038/s41561-021-00812-2

[83] Cayuela ML, van Zwieten L, Singh BP, et al. Biochar's role in mitigating soil nitrous oxide emissions: A review and meta-analysis. Global Change Biology Bioenergy. 2014;**6**(4):401-415. DOI: 10.1111/gcbb.12022

[84] Luo X, Feng H, Liu J, et al. Meta-analysis on farmland soil N_2O emissions under biochar application in China. Chinese Journal of Eco-Agriculture. 2017;**25**(9):1254-1265. DOI: 10.13930/j.cnki.cjea.170298

[85] Jeffery S, Verheijen FGA, van der Velde M, et al. A quantitative review of the effects of biochar application to soils on crop productivity using meta-analysis. Global Change Biology Bioenergy. 2016;**8**(5):960-976. DOI: 10.1111/gcbb.12285

[86] Zhang Y, Wang H, Lei Q, et al. Optimizing the nitrogen application rate for maize and wheat based on yield and environment on the Northern China Plain. The Science of the Total Environment. 2018;**618**(MAR.15):1173-1183. DOI: 10.1016/j.scitotenv.2017.09.183

[87] Akhtar SS, Li G, Andersen MN, et al. Biochar mitigates salinity stress in potato. Journal of Agronomy and Crop Science. 2014;**200**(5):371-380. DOI: 10.1111/jac.12070

[88] Burrell LD, Zehetner F, Rampazzo N, et al. Long-term effects of biochar on soil physical properties. Geoderma: An International Journal of Soil Science. 2016;**282**:96-102. DOI: 10.1016/j.geoderma.2016.07.019

[89] Major JM, Rondón Molina D, et al. Maize yield and nutrition during 4years after biochar application to a Colombian savanna oxiso. Plant and Soil. 2010;**333**:117-128. DOI: 10.1007/s11104-010-0327-0

[90] Zhang A, Bian R, Pan G, et al. Effects of biochar amendment on soil quality, crop yield and greenhouse gas emission in a Chinese rice paddy: A field study of 2 consecutive rice growing cycles. Field Crops Research. 2012;**127**:153-160. DOI: 10.1016/j.fcr.2011.11.020

[91] Liu X, Zhang A, Ji C, et al. Biochar's effect on crop productivity and the dependence on experimental conditions—A meta-analysis of literature data. Plant and Soil. 2013;**373**:583-594. DOI: 10.1007/s11104-013-1806-x

[92] Bonanomi G, Antignani V, Capodilupo M, et al. Identifying the characteristics of organic soil amendments that suppress soilborne plant diseases. Soil Biology and Biochemistry. 2010;**42**(2):136-144. DOI: 10.1016/j.soilbio.2009.10.012

[93] Zheng ALT, Lih ETY, Hung YP, et al. Biochar-based electrochemical sensors: A tailored approach to environmental monitoring. Analytical Sciences. 2025;**41**(6):715-735. DOI: 10.1007/s44211-025-00726-8

[94] Zhang A, Zhou X, Li M, et al. Impacts of biochar addition on soil dissolved organic matter characteristics in a wheat-maize rotation system in Loess Plateau of China. Chemosphere. 2017;**186**:986. DOI: 10.1016/j.chemosphere.2017.08.074

[95] Downie A, Munroe P, Cowie A, et al. Biochar as a geoengineering climate solution: Hazard identification and risk management. Critical Reviews in Environmental Science and Technology. 2012;**42**(3):225-250. DOI: 10.1080/10643389.2010.507980

[96] Australian Government, Clean Energy Regulator. Available from: https://www.sciencedirect.com/science/article/pii/S0960148113003170.pdf [Accessed: June 24, 2013]

[97] Ippolito JA, Cui L, Kammann C, et al. Feedstock choice, pyrolysis temperature and type influence biochar characteristics: A comprehensive meta-data analysis review. Biochar. 2020;**2**:421-438. DOI: 10.1007/s42773-020-00067-x

[98] Jaffari ZH, Jeong H, Shin J, et al. Machine-learning-based prediction and optimization of emerging contaminants' adsorption capacity on biochar materials. Chemical Engineering Journal. 2023;**466**(000):12. DOI: 10.1016/j.cej.2023.143073

[99] Pathy A, Balasubramanian P. Predicting algal biochar yield using eXtreme Gradient Boosting (XGB) algorithm of machine learning methods. Algal Research. 2020;**50**(September 2020):102006. DOI: 10.1016/j.algal.2020.102006

[100] Purcell W, Neubauer T. Digital twins in agriculture: A state-of-the-art review. Smart Agricultural Technology. 2023;**3**:100094. DOI: 10.1016/j.atech.2022.100094.4

[101] Mohammed AKO, Alam A, Hotak Y, et al. Smart sensor technologies shaping the future of precision agriculture: Recent advances and future outlooks. Journal of Sensors. 2025;**2025**:2460098. DOI: 10.1155/js/2460098

[102] Zhang Z, Schott JA, Liu M, et al. Prediction of carbon dioxide adsorption via deep learning. Angewandte Chemie International Edition. 2019;**58**:259. DOI: 10.1002/anie.201812363

[103] Huber R, Xiong H, Keller K, et al. Bridging behavioural factors and standard bio-economic modelling in an agent-based modelling framework. Journal of Agricultural Economics. 2022;**73**(1):35-63. DOI: 10.1111/1477-9552.12447

[104] Scholz SM, Sembres T, Roberts K, et al. Biochar Systems for Smallholders in Developing Countries: Leveraging Current Knowledge and Exploring Future Potential for Climate-Smart Agriculture. Washington, DC, USA: World Bank Publications; 2014. DOI: 10.1596/978-0-8213-9525-7

[105] Markusson N. Political obstacles to carbon capture and storage for carbon removal. Nature Reviews Earth & Environment. 2024;**5**(7):481-482. DOI: 10.1038/s43017-024-00574-6

[106] Whitford AB, Anderson D. Governance landscapes for emerging technologies: The case of cryptocurrencies. Regulation & Governance. 2021;**15**(4):1053-1070. DOI: 10.1111/rego.12366

[107] Bednar J, Obersteiner M, Baklanov A, et al. Operationalizing the net-negative carbon economy. Nature. 2021;**596**:377-383. DOI:10.1038/s41586-021-03723-9

[108] Günther P, Garske B, Heyl K, et al. Carbon farming, overestimated negative emissions and the limits to emissions trading in land-use governance: The EU carbon removal certification proposal. Environmental Sciences Europe. 2024;**36**(1):72. DOI: 10.1186/s12302-024-00892-y

Chapter 2

Combined Thermochemical Conversion as a Pretreatment of Manure Waste for Environmental Applications: An Overview

Kalidas Mainali

Abstract

Globally, significant amounts of manure waste are generated. The primary method for valorizing manure waste is through its terrestrial use. This practice can lead to various environmental issues, including the spread of infections, the release of unpleasant substances (such as ammonia or methane), and the contamination of aquatic systems with nitrogen and phosphorus compounds. However, the cellulosic substrates of manure biogas yield can be improved by combined thermal treatment along with wet oxidation. In fact, integrating mild combined treatment prior to anaerobic digestion maximizes methane yield. The solid digestate could serve as a renewable source for N-doped chars. Alternatively, a new carbonization technique can concentrate all nutrients on the solid product. These compacted solid products are rich in nitrogen and phosphorus, making their transportation over long distances feasible for eventual use as soil amendments. Enhancing the carbon and nitrogen content in biochar derived from raw dairy manure and solid digestate through thermochemical methods may promote sustainable manure consumption and management. The biochar produced from large quantities of dairy waste serves as a tool for environmental sustainability.

Keywords: manure waste, thermal treatment, environmental issues, biochar, pyrolysis, carbonization

1. Introduction

Our world is facing a growing array of global risks and environmental challenges, including climate change, clean water scarcity, and environmental contamination. With the increasing global population, the demand for food resources is escalating [1, 2]. Livestock production has undergone significant transformation over the past few decades. Predominantly, meat and dairy products are produced on large farms characterized by single-species operations, leading to a twofold increase in milk output, a threefold increase in meat production, and a fourfold increase in egg production [3, 4]. While concentrated animal feeding operations (CAFOs) offer benefits such as economical sources of meat, milk, and eggs, they also present significant challenges.

Manure and wastewater contain numerous potential pollutants, including plant nutrients like nitrogen and phosphorus, pathogens, antibiotics, and other chemicals. Effectively managing and addressing wastewater pollution concerns is vital for protecting health, ensuring food quality, preserving the environment, creating employment opportunities, and generating energy [5–7].

The intensive CAFO system, a vital part of the agricultural economy, produces large amounts of solid organic waste, which creates ecological challenges and public health risks [8–10]. The use of biological processes for solid waste treatment is gaining increased interest. Anaerobic digestion (AD) is recognized as an energy- efficient method for converting organic waste into energy. It offers a unique therapeutic approach for animal husbandry, providing benefits related to several issues, including renewable energy [11, 12]. AD involves a series of metabolic reactions that mainly occur without oxygen, proceeding through four stages: hydrolysis, acidogenesis, acetogenesis, and methanogenesis [13]. Hydrolysis is the first step, where bacteria use enzymes to break down high- molecular- weight substances into soluble molecules or monomers. However, the stubborn nature of manure limits microbial access to key components—cellulose and hemicellulose—causing a rate- limiting step. The digestibility of cellulose in lignocellulosic biomass is hindered by various physical, chemical, structural, and compositional factors. Studies show that only 20–30% of the initial biomass converts to methane during anaerobic digestion, with large amounts of carbon remaining as cellulose, hemicellulose, lignin, and proteins [14]. Additionally, improper handling of large amounts of digestate could lead to significant environmental issues. Converting these residual materials into biofuel is complex. Lignin, among these residues, is generally the most difficult to digest. Moreover, hydrogen bonds within the hydroxyl groups of cellulose strengthen a supramolecular structure with high crystallinity, further complicating digestion. Overall, these lignocellulosic materials greatly impact the hydrolysis process, forming a barrier that hinders microbial activity in substrate breakdown. Therefore, improving the digestibility of this feedstock is crucial to increase overall methane production. Meanwhile, mild hydrothermal pretreatment before anaerobic digestion is necessary to break down lignocellulosic fibers.

Prior research indicates that hydrothermal treatment eliminates the majority of hemicellulose from raw materials [11, 15]. Nonetheless, it is less efficacious in eliminating the lignin and other constituents. Due to its composition of 40–50% fibrous lignocellulosic material, dairy manure is not significantly amenable to lignin breakdown with hydrothermal processing alone [7]. Wet oxidation is a key technology mostly used in the disposal of industrial and agricultural wastewater. However, it has received very little attention for treating solid waste (such as dairy manure). Along with thermal hydrolysis, wet oxidation is widely used to treat organic waste effluents. Wet oxidation has advantages over other alternative treatment technologies as it avoids the need for a dewatering step [7, 16]. Consequently, the integration of hydrothermal pretreatment with alkali supplementation seems to be an efficacious approach for lignin breakdown. Moreover, the application of ammonia enhances alkalinity with low input, necessitating reduced chemical preparation relative to acidic treatment [17]. This results from the nitrogenous content in manure, which increases alkalinity. The utilization of ammonia generates fewer inhibitors, exhibits reduced corrosiveness to reactors, and allows for straightforward recycling. The efficient application of combined hydrothermal treatment prior to the anaerobic digestion (AD) process may offer a sustainable method for the usage of dairy manure.

The main goal of this chapter is to offer a comprehensive overview of combined hydrothermal pretreatment and wet oxidation of dairy manure for a sustainable anaerobic biorefinery. Applying mild hydrothermal and wet oxidation treatments before anaerobic digestion can support sustainable dairy manure management with renewable energy. To fill the knowledge gap between anaerobic digestion and manure pretreatment, it is suggested that hydrothermal processing combined with wet oxidation be used as a pretreatment method to improve manure lignocellulosic breakdown.

2. Enhancement of anaerobic digestion of animal manure for biogas production

The generation of biogas from lignocellulosic materials offers a significant opportunity to convert waste manure into renewable energy [8, 18]. Despite its substantial methane potential, manure has not been efficiently used for biogas production. The main challenge with manure as a substrate is its effective digestion during the anaerobic digestion (AD) process, due to the complex and resistant nature of the feedstock [2, 10, 11]. Manure contains polymers such as cellulose, hemicellulose, and lignin; cellulose is the primary component of lignocellulosic material, surrounded by a hemicellulose layer and an outer lignin layer. The literature describes several pretreatment methods, including physical, chemical, and biological techniques, for treating lignocellulosic biomass, organic waste, sewage sludge, agro-waste, and other materials. However, there is limited research on preparing animal manure, mainly because its high fiber content is a significant barrier to effective utilization in anaerobic digestion. The main goal of pretreatment is to break the bonds between lignin and hemicellulose, reduce the crystallinity of cellulose, and increase the porosity of lignocellulosic materials. Various pretreatment methods have been studied to achieve these aims. Pretreatment is a common approach used to improve both the rate and yield of methane production [2, 10, 19]. **Figure 1** below shows BMP results after pretreating the specified substrates.

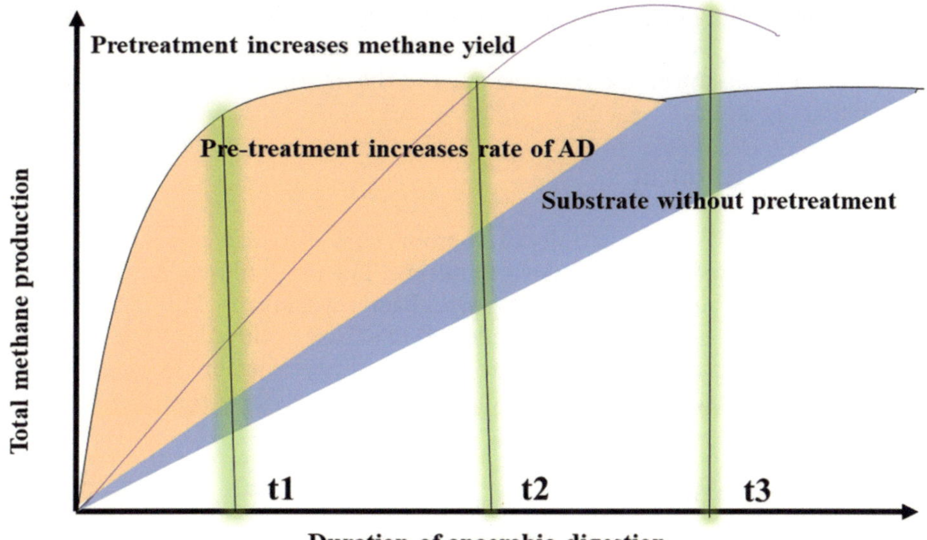

Figure 1.
Conceptual diagram of the effect of pretreatment on biogas yield (Redrawn, [20]).

3. Overview of the pretreatment procedure

3.1 Physical pretreatment

It can be classified into mechanical and thermal treatments. Mechanical pretreatments encompass milling, extrusion, grinding, and similar processes. The physical properties of biomass (manure) are altered using screens, sieves, presses, and centrifuges. Maceration is a frequently employed physical treatment method for animal manure. This procedure enhances the fiber's surface area, facilitating microbial access to the substrate and augmenting methane output. In literature, mechanical separation is extensively employed for dairy manure utilizing devices such as screens, presses, and centrifuges. Masse et al. [21] developed a solid-liquid separation system that effectively eliminates a substantial fraction of solids to diminish phosphorus and heavy metal concentrations in raw manure [21]. Likewise, literature indicates that thermal pretreatment involves the gradual heating of biomass to release volatiles and alter its structure, hence enhancing digestibility [22]. This approach is energy-intensive, costly, and impractical for bio-refinery applications.

3.2 Chemical pretreatment

This involves the application of chemicals (including acids, bases, and organic solvents) to modify the physical and chemical properties of lignocellulose. It has been extensively examined in literature due to its efficacy in reducing stubborn traits. Acid pretreatment effectively hydrolyzes hemicellulose, hence augmenting the pore size or volume of the cell wall, which renders cellulose more amenable to enzymatic breakdown [23]. It also interferes with the lignin structure. The application of diluted and concentrated acid in the hydrolysis of lignocellulosic biomass improved sugar output. Alkaline pretreatment involves the solvation of lignocellulose particles and the hydrolytic degradation of lignocellulose through saponification. This technique improves hemicellulose dissolution, augments internal surface area, and decreases polymerization [24]. Nonetheless, they possess disadvantages like the elevated expense of acids, susceptibility to corrosion, and the necessity for neutralization prior to subsequent digestion.

3.3 Biological pretreatment

This technique utilizes the activity of microorganisms to alter the chemical composition and structure of lignocellulosic materials. The predominant microorganisms are white and brown soft-rot fungi, along with bacteria. Biological pretreatments are promising technologies due to their chemical-free nature, reduced energy use, and environmental sustainability. Nonetheless, biological pretreatment is less efficient, more time-consuming, and requires a regulated environment for optimal effectiveness [25].

3.4 Hydrothermal carbonization

(HTC) is a thermochemical conversion process that employs lignocellulosic biomass to generate a solid, coal-like material. Introduced by Bergius in 1913, it replicates the natural coal formation process wherein cellulose converts into coal-like substances in a laboratory environment [13, 26]. HTC presents advantages over other thermochemical processes, including the capacity to utilize biomass with elevated moisture content directly. The HTC process generally takes place at low temperatures (180–250°C) and under autogenous pressure, rendering it especially appropriates for wet materials, such as raw dairy manure and sludge, which comprises approximately 80% dairy manure [10, 27].

The reaction mechanism of hydrothermal carbonization has been extensively examined in literature. The entire process is regulated by dehydration and decarboxylation. Initially, hydrolysis events result in the breaking of mostly ester and ether bonds within macromolecules with the addition of water [26]. This procedure yielded a diverse array of compounds, including cellulose oligosaccharides and phenolic lignin fragments. The dehydration process encompasses both chemical and physical mechanisms via which water is extracted from the biomass matrix. Chemical dehydration markedly carbonizes biomass by reducing the H/C and O/C ratios through the removal of hydroxyl groups. General hydrothermal treatment results in a partial removal of carboxyl groups. The response mechanisms remain mainly unidentified, particularly the influence of water's presence. The carboxyl and carbonyl groups swiftly decompose above 150°C, producing CO_2 and CO, respectively [28–30]. Probable sources of CO_2 include short-chain acids, specifically formic and acetic acids, which are produced in substantial quantities during the decomposition of cellulose and hemicellulose. Additionally, there are condensation processes and the breaking of intramolecular bonds. Moreover, H_2O functions as an oxidizing agent at high temperatures, resulting in the formation of extra CO_2 through the thermal decomposition of these oxidized units. Ultimately, the fragments of biomolecule degradation aggregate, leading to polymerization. The creation of HTC products is primarily characterized by condensation polymerization, particularly aldol condensation. Condensation predominantly occurs with the generation of water. The mechanism of free radicals may also occur in hydrothermal circumstances, but it is likely to be predominant at supercritical conditions and at low density.

Agricultural products are produced in substantial quantities worldwide each year. It presents significant potential to transform these plentiful resources into energy and valued products. The most promising technologies for converting trash into biofuels are hydrothermal procedures that utilize subcritical and supercritical water as processing media. Wilk et al. [31] conducted comparative analyses of sewage sludge and lignocellulosic biomass using hydrothermal carbonization [31]. A substantial increase in fixed carbon was seen in the case of lignocellulosic biomass. Conversely, in sludges, carbon compounds transitioned to the liquid phases, resulting in diminished carbonaceous products. In additional investigations, an anaerobically digested sludge was carbonized using hydrothermal carbonization to provide clean solid fuels. The hydrothermal carbonization of cow dung was also examined in the literature [32]. This investigation involved the hydrothermal carbonization of manure at temperatures ranging from 180 to 260°C for durations of 5 and 30 minutes. The carbon content in hydro-char escalates with temperature, and the higher heating value (HHV) of hydro-char was documented as 22.1 MJ/kg [32]. Hydro-chars can serve as adsorbents for the remediation of environmental contaminants in water, particularly when their surfaces are functionalized with specific oxygen and nitrogen groups. Currently, carbon compounds modified with nitrogen, sulfur, and phosphorus groups are under extensive investigation as carbon electrodes for supercapacitors. Lentz et al. [33] investigated swine dung slurry using hydrothermal carbonization and reported a fuel with high energy content [33].

3.5 Wet air oxidation

The wet oxidation technique was initially marketed for the production of artificial vanilla flavoring and subsequently employed for the degradation of paper-mill sludge and biological sludge [34]. This process involves the oxidation of organic and inorganic compounds in an aqueous solution using oxygen or air under increased temperatures and pressures. The primary processes resemble incineration (oxidation) in water through WAO. The WAO method is optimal for the treatment of waste liquors, slurries, and similar materials with a high concentration of organic matter. Consequently, this technique is extensively employed to cleanse organic effluents and wastewater. This is a form of hydrothermal treatment for the removal of particulates in organic waste. Prior research demonstrated that a 90% reduction in total suspended solids and volatile suspended solids may be attained by the wet oxidation procedure [35, 36]. Research has examined the diverse organic chemicals, industrial processes, and wastewater through wet oxidation and catalytic wet air oxidation [35, 37, 38]. Nonetheless, leaching of metal catalysts (e.g., CuO) and catalyst fouling were reported during the catalytic wet oxidation process [36]. The fundamental mechanism of the moist air oxidation process entails the complete oxidation of organic molecules into carbon dioxide and water under conditions of elevated temperature and pressure. Under elevated temperature and pressure, molecular oxygen begins to solubilize in the aqueous medium and interacts with organic compounds, yielding CO_2 and H_2O through organic acid intermediates [7]. In this process, water constitutes the predominant aqueous phase and functions as an efficient heat transfer medium in the presence of Air/O_2 to alter oxidation processes. In the primary chain reaction, free radicals are generated through the bimolecular interaction of oxygen with organic molecules [16, 39]. Considering all these scenarios, introducing wet oxidation along with hydrothermal as a mild treatment for dairy manure works as an effective treatment. The procedure is devoid of any deleterious chemicals or reagents. In conclusion, mild wet oxidation treatment of dairy manure enhances solubility as well as deconstructs lignocellulosic fiber of manure, which enhances methane yield [7].

Anaerobic digestion is a widely utilized manure method [18]. The primary obstacle in utilizing dairy manure as a substrate for anaerobic digestion is the low bio-digestibility of its fiber content [12]. Holm-Nielson et al. [40] revealed that animal dung has low biodegradability in anaerobic digestion due to its high fiber content [40]. Various hydrothermal processing strategies have been employed in the literature for the efficient exploitation of manure, including thermal hydrolysis (TH), hydrothermal carbonization (HTC), and hydrothermal liquefaction (HTL). Nonetheless, each process generates certain inhibitory chemicals, including furfural, 5-HMF, phenol, and phenolic derivatives. Furthermore, HTC and HTL necessitate extremely elevated temperature and pressure conditions. Conversely, conventional biochemical processes like fermentation can convert biomass into volatile fatty acids; however, the reaction rates are sluggish and require several days to complete. Wet air oxidation of dairy manure has garnered significant attention and can be accomplished in a matter of minutes.

4. Selective carbonization of nitrogen-infused biomass

Carbonization is the process whereby high-carbon-content solid residues are produced from organic material, often through pyrolysis in an inert atmosphere [26]. Slow pyrolysis is primarily used to achieve optimal char yield. Numerous studies have explored the impact of pyrolysis operational parameters on the yield of products derived from lignocellulosic biomass. The kinetic and thermodynamic behavior of lignocellulosic materials during pyrolysis has been investigated to enhance understanding of the decomposition stages. Pyrolysis is the main method used for producing biochar and bio-oil [26]. The solid product serves various purposes, while bio-oil can function as fuel and be refined to produce chemicals. Despite the high energy content of bio-oil derived from biomass pyrolysis, it has undesirable characteristics, including high viscosity and corrosiveness. Previously, research focused on optimizing bio-oil production. The pyrolysis of protein-rich biomass (such as sewage sludge, dairy manure, and energy crops) presents a significant challenge due to the high nitrogen concentration in bio-oil, which leads to NOx emissions. Utilizing bio-crude as fuel requires additional technical measures to control nitrogen emissions to comply with fuel regulations [26, 41].

Numerous investigations have been undertaken in the literature regarding the slow and quick pyrolysis of protein-dense biomass, including sewage sludge [41]. This study indicates that the nitrogen content of biooil at 400°C was reported to be between 8.4 and 9.3 wt. %. Furthermore, the nitrogen content of bio-oil may attain up to 16 weight percent. Percentage during the catalytic pyrolysis of Spirulina (nitrogen-rich biomass). High nitrogen-content bio-oils would substantially impede the utilization of bio-oil as fuel. In a similar vein, the hydrothermal liquefaction of biomass with varying nitrogen content and model compounds was thoroughly investigated to ascertain the behavior of nitrogen and nitrogen-containing heterocycles in the biocrude [26, 41]. The results indicated that the nitrogen concentration in bio-oil increased with elevated reaction temperatures in actual feedstocks. Furthermore, numerous investigations have suggested that Millard reactions, which include the browning reaction between reducing sugars and amino groups at elevated temperatures, affect thermal hydrolysis. Nitrogen-containing recalcitrant organics may be synthesized via Millard reactions [26]. The chemistry of the Maillard reaction is intricate owing to numerous reaction pathways, such as condensation, cyclization, dehydration, and rearrangements. Nevertheless, there is a paucity of research papers elucidating the impact of Millard responses in relation to dairy manure.

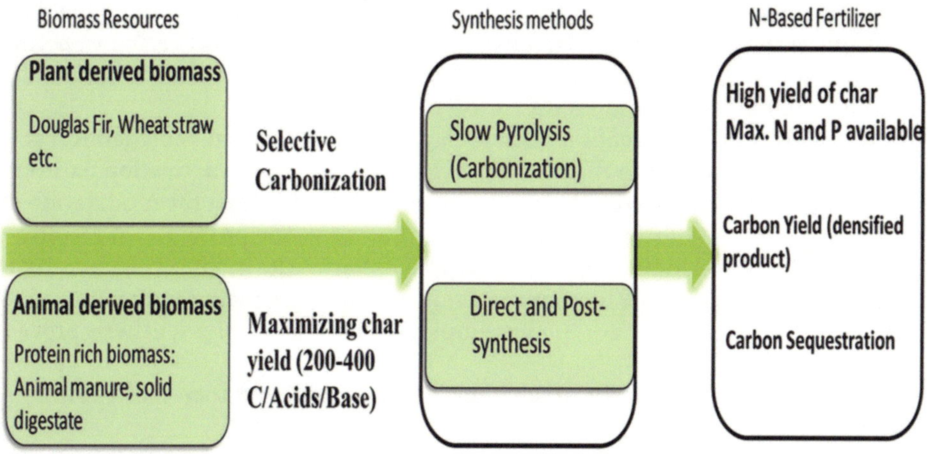

Figure 2.
Selective carbonization to maximize carbon yield and enrich in N biochar structure.

The transition behavior of nitrogen during pyrolysis remains inadequately understood. The impacts of biomass type (N/C ratio) and biochemical processes on mineralization are inadequately comprehended. Consequently, optimizing solid carbon output and the retention of nitrogen and phosphorus in the solid phase may provide a novel approach for the sustainable utilization of manure. The utilization of acids and other desiccants may facilitate the catalysis and removal of oxygen in the form of water. The pyrolysis of acid-impregnated lignocellulosic biomass has been examined in the literature [10]. The pyrolysis of lignocellulosic biomass was mostly influenced by dehydration reactions rather than depolymerization reactions. Given these considerations, the investigation of a novel carbonization approach (selective carbonization) should prioritize future thermochemical processes. **Figure 2** illustrates the proposed selective carbonization process utilizing bioresources, including dairy manure.

5. Biochar derived from manure and its usage in environmental contexts

Despite the long-standing establishment of biogas production via anaerobic digestion, optimization of this process is necessary to enhance stability, increase methane outputs, and address inhibitory challenges [42]. A prominent issue related to the anaerobic digestion of nitrogen-laden materials is the inhibition caused by excessive ammonia. Conversely, the nitrogen-rich solid digestate may serve as a renewable feedstock for biochar. The solid component of digestate can serve as bedding material or soil amendments, albeit alternatives are restricted. The elevated organic matter and mineral nutrient content render animal manure an accessible fuel for bioenergy and biochar synthesis. Conversely, the substantial waste of amoebic-digested fiber, abundant in nitrogen and phosphorus, can function as an environmental resource. In recent years, biochar has increasingly garnered attention. Biochar is a carbon-dense solid produced by pyrolyzing biomass, including wood or manure, in an oxygen-limited atmosphere. It can persist in soil for millennia [43]. As biochar research progresses, its significant environmental and economic advantages are widely acknowledged, prompting scientific investigations for its application in engineering. The upcycling of extensive manure waste by thermochemical and biological transformations into sophisticated carbon materials offers a viable alternative. While biochar research is compelling, numerous problems remain to be resolved for dependable applications. Manure typically exhibits complicated compositions, varied structures, and a heterogeneous character, posing considerable obstacles for the standardization of established preparation and modification processes [44].

A diverse array of modification techniques, including physical and chemical activation, has been documented in the literature [26, 45]. The chosen activation pathways and their application of diverse chemicals can substantially affect the modulation of pore architectures and biochar production. Furthermore, the protein-rich manure may function as an in-situ heteroatom dopant to enhance the intrinsic characteristics of biochar. While these waste biomass feedstocks inherently include advantageous dopants, external precursors are further employed throughout the synthesis process.

5.1 Biochar serves as a versatile adsorbent, catalyst, soil enhancer, and energy storage medium

Biochar is cost-effective and may be readily produced from various organic waste materials, including dairy manure, through thermal methods. By optimizing its distinctive characteristics, including pore volume, surface area, long-term stability, and enhanced functional groups, biochar can be utilized in various applications such as wastewater treatment, catalysis, soil enhancement, fuel production, carbon sequestration, and energy storage [26]. Biochar is regarded as a promising universal adsorbent for the elimination of diverse contaminants in wastewater and soils. Biochar can adsorb heavy metals, organic pollutants, nitrogen, phosphorus toxins, PFAS, and various other contaminants. The physicochemical characteristics of the specified biochar govern the adsorption process. Engineering chars with specialized features can be designed by physical and chemical activation to target various pollutants and impurities. The plentiful surface groups, primarily oxygen-containing groups like -COOH and -OH, exhibit a robust interaction with heavy metals, hence facilitating the adsorption process. Furthermore, the surface area and porous architecture of biochar influence the adsorption of heavy metals. In agricultural settings, the occurrence of per- and polyfluoroalkyl substances (PFASs) in water may endanger irrigated crops and the human food supply [46]. Biochar may serve as a cost-effective and eco-friendly method for the removal of PFAS from water. Ramos et al. [47] conducted a thorough assessment of several biochars derived from agricultural waste, including manure, and determined that, in addition to surface area, the carbon and nitrogen content of designed chars were effective in eliminating both long-chain and short-chain PFAS [47].

Biochar is regarded as a promising support material or catalyst used in environmental and energy applications. Its comparatively high specific surface area and pore volume enhance catalytic activities by providing additional active sites for organic reactions. Furthermore, metal-enriched biochar feedstocks may function as support materials in catalysis. Dispersed findings indicate that hardwood-derived biochar demonstrates significant activity in esterification. Extensive literature supports biochar as an electrode material in supercapacitors [26].

Soil is crucial in terrestrial ecosystems. Soil degradation has occurred in recent decades due to poor land management, deforestation, and climate change [48]. These practices pose considerable risks to food security and agricultural sustainability. The unique properties of biochar, including its large surface area, stable carbon composition, and high nutritional content, make it an appealing candidate for soil rehabilitation. Soil remediation with biochar primarily happens through adsorption. The biochar adsorption mechanism includes surface complexation, hydrogen bonding, and electrostatic attraction, among others. Biochar is reported to effectively remove organic contaminants and heavy metals, enhance cation exchange capacity, and improve soil fertility [49]. Moreover, soil, as a key carbon sink, plays an important role in the global carbon cycle. Due to its highly condensed aromatic structure, biochar shows significant resistance to biodegradation, thus positively influencing carbon sequestration in the soil.

6. Conclusions

Livestock production is a vital part of the global agricultural system. Excessive manure application to land causes social, economic, and environmental issues. Pretreating manure is an alternative method that improves anaerobic decomposition. The remaining solid digestate can be used as a renewable feedstock for biochar production. Combining hydrothermal treatment before anaerobic digestion can reduce greenhouse gases, increase methane yield, and create a renewable energy source.

Additionally, this integrated approach offers significant benefits for sustainable manure waste management, with clean energy production as an added advantage. Making biochar through selective carbonization from dairy farms is an eco-friendly method. The protein-rich biochar can positively impact rural economies and serve as a solution for environmental restoration. Converting inexpensive manure feedstocks into biochar addresses waste management issues and supports environmental rehabilitation. These combined techniques, along with anaerobic digestion, help lower pollutants and greenhouse gas emissions, improve nutrient recovery, and boost bioenergy production, fostering a circular bioeconomy.

Author details

Kalidas Mainali[1,2]

1 US Department of Agriculture, Agricultural Research Service, Eastern Regional Research Center, Sustainable Biofuels and Co-Products Research Unit, Wyndmoor, PA, USA

2 Biological Systems Engineering, Washington State University, Pullman, WA, USA

*Address all correspondence to: kalidas.mainali@usda.gov

References

[1] Bradford KJ, Dahal P, Van Asbrouck J, Kunusoth K, Bello P, Thompson J, et al. The dry chain: Reducing postharvest losses and improving food safety in humid climates. In: Food Industry Wastes. Academic Press; 2020. pp. 375-389

[2] Orlando M-Q, Borja V-M. Pretreatment of animal manure biomass to improve biogas production: A review. Energies. 2020;**13**(14):3573

[3] Cole D, Todd L, Wing S. Concentrated swine feeding operations and public health: A review of occupational and community health effects. Environmental Health Perspectives. 2000;**108**(8):685-699

[4] Dadrasnia A, de Bona I, Muñoz EH, Yáñez IU, Lamkaddam MM, Ponsá S, et al. Sustainable nutrient recovery from animal manure: A review of current best practice technology and the potential for freeze concentration. Journal of Cleaner Production. 2021;**315**:128106

[5] Fermoso FG, Serrano A, Alonso-Fariñas B, Fernández-Bolaños J, Borja R, Rodríguez-Gutiérrez G. Valuable compound extraction, anaerobic digestion, and composting: A leading biorefinery approach for agricultural wastes. Journal of Agricultural and Food Chemistry. 2018;**66**(32):8451-8468

[6] Hribar C. Understanding Concentrated Animal Feeding Operations and their Impact on Communities. 2010

[7] Mainali K, Mood SH, Chen S, Garcia-Perez M. Partial wet oxidation of dairy manure as a pretreatment process to produce acetic acid 'A source growth of methanogens'. Waste Management & Research. 2024;**42**(3):206-217

[8] Awasthi MK, Sarsaiya S, Wainaina S, Rajendran K, Kumar S, Quan W, et al. A critical review of organic manure biorefinery models toward sustainable circular bioeconomy: Technological challenges, advancements, innovations, and future perspectives. Renewable and Sustainable Energy Reviews. 2019;**111**:115-131

[9] He Z, Pagliari P, Waldrip HM. Advances and outlook of manure production and management. Animal Manure: Production, Characteristics, Environmental Concerns, and Management. 2020;**67**:373-383

[10] Mainali K, Sarker MI, Mullen CA, Sharma BK, Yadav MP, Ngo H, et al. Thermal decomposition kinetics of dairy manure hydrochars. Journal of the Energy Institute. 2025;**120**:102088

[11] Ahmad F, Silva EL, Varesche MBA. Hydrothermal processing of biomass for anaerobic digestion–a review. Renewable and Sustainable Energy Reviews. 2018;**98**:108-124

[12] Nasir M, Ismail, Ghazi TIM. Pretreatment of lignocellulosic biomass from animal manure as a means of enhancing biogas production. Engineering in Life Sciences. 2015;**15**(7):733-742

[13] Gong L, Yang X, Wang Z, Zhou J, You X. Impact of hydrothermal pre-treatment on the anaerobic digestion of different solid–liquid ratio sludges and kinetic analysis. RSC Advances. 2019;**9**(33):19104-19113

[14] Xu N, Liu S, Xin F, Zhou J, Jia H, Jiming X, et al. Biomethane production from lignocellulose: Biomass recalcitrance and its impacts

on anaerobic digestion. Frontiers in Bioengineering and Biotechnology. 2019;**7**:191

[15] Scapini T, Dos Santos MSN, Bonatto C, Wancura JHC, Mulinari J, Camargo AF, et al. Hydrothermal pretreatment of lignocellulosic biomass for hemicellulose recovery. Bioresource Technology. 2021;**342**:126033

[16] Rivas FJ, Kolaczkowski ST, Beltran FJ, McLurgh DB. Hydrogen peroxide promoted wet air oxidation of phenol: Influence of operating conditions and homogeneous metal catalysts. Journal of Chemical Technology & Biotechnology. 1999;**74**(5):390-398

[17] Kim JS, Lee YY, Kim TH. A review on alkaline pretreatment technology for bioconversion of lignocellulosic biomass. Bioresource Technology. 2016;**199**:42-48

[18] Meegoda JN, Li B, Patel K, Wang LB. A review of the processes, parameters, and optimization of anaerobic digestion. International Journal of Environmental Research and Public Health. 2018;**15**(10):2224

[19] Chen Y, Ping Q, Li D, Dai X, Li Y. Comprehensive insights into the impact of pretreatment on anaerobic digestion of waste active sludge from perspectives of organic matter composition, thermodynamics, and multi-omics. Water Research. 2022;**226**:119240

[20] Atelge MR, Atabani AE, Rajesh Banu J, David Krisa M, Kaya CE, Kumar G, et al. A critical review of pretreatment technologies to enhance anaerobic digestion and energy recovery. Fuel. 2020;**270**:117494

[21] Masse L, Massé DI, Beaudette V, Muir M. Size distribution and composition of particles in raw and anaerobically digested swine manure. Transactions of the ASAE. 2005;**48**(5):1943-1949

[22] Karthikeyan PK, Bandulasena HCH, Radu T. A comparative analysis of pre-treatment technologies for enhanced biogas production from anaerobic digestion of lignocellulosic waste. Industrial Crops and Products. 2024;**215**:118591

[23] Yin S, Tan Z. Hydrothermal liquefaction of cellulose to bio-oil under acidic, neutral and alkaline conditions. Applied Energy. 2012;**92**:234-239

[24] Perendeci NA, Ciggin AS, Kökdemir Ünşar E, Orhon D. Optimization of alkaline hydrothermal pretreatment of biological sludge for enhanced methane generation under anaerobic conditions. Waste Management. 2020;**107**:9-19

[25] Awasthi MK, Sar T, Gowd SC, Rajendran K, Kumar V, Sarsaiya S, et al. A comprehensive review on thermochemical, and biochemical conversion methods of lignocellulosic biomass into valuable end product. Fuel. 2023;**342**:127790

[26] Mainali K, Mood SH, Pelaez-Samaniego MR, Sierra-Jimenez V, Garcia-Perez M. Production and applications of N-doped carbons from bioresources: A review. Catalysis Today. 2023;**423**:114248

[27] Song C, Zheng H, Shan S, Shengchun W, Wang H, Christie P. Low-temperature hydrothermal carbonization of fresh pig manure: Effects of temperature on characteristics of hydrochars. Journal of Environmental Engineering. 2019;**145**(6):04019029

[28] Kim D, Lee K, Park KY. Hydrothermal carbonization of anaerobically digested sludge for solid

fuel production and energy recovery. Fuel. 2014;**130**:120-125

[29] Wu K, Gao Y, Zhu G, Zhu J, Yuan Q, Chen Y, et al. Characterization of dairy manure hydrochar and aqueous phase products generated by hydrothermal carbonization at different temperatures. Journal of Analytical and Applied Pyrolysis. 2017;**127**:335-342

[30] Xiao L-P, Shi Z-J, Feng X, Sun R-C. Hydrothermal carbonization of lignocellulosic biomass. Bioresource Technology. 2012;**118**:619-623

[31] Wilk M, Magdziarz A, Jayaraman K, Szymańska-Chargot M, Gökalp I. Hydrothermal carbonization characteristics of sewage sludge and lignocellulosic biomass. A comparative study. Biomass and Bioenergy. 2019;**120**:166-175

[32] Toufiq Reza M, Freitas A, Yang X, Hiibel S, Lin H, Coronella CJ. Hydrothermal carbonization (HTC) of cow manure: Carbon and nitrogen distributions in HTC products. Environmental Progress & Sustainable Energy. 2016;**35**(4):1002-1011

[33] Lentz Z, Kolar P, Classen J. Valorization of swine manure into Hydrochars. PRO. 2019;7:560

[34] Hii K, Baroutian S, Parthasarathy R, Gapes DJ, Eshtiaghi N. A review of wet air oxidation and thermal hydrolysis technologies in sludge treatment. Bioresource Technology. 2014;**155**:289-299

[35] Chung J, Lee M, Ahn J, Bae W, Lee Y-W, Shim H. Effects of operational conditions on sludge degradation and organic acids formation in low-critical wet air oxidation. Journal of Hazardous Materials. 2009;**162**(1):10-16

[36] Reza MT, Freitas A, Yang X, Coronella CJ. Wet air oxidation of hydrothermal carbonization (HTC) process liquid. ACS Sustainable Chemistry & Engineering. 2016;**4**(6):3250-3254

[37] Arena F, Di Chio R, Gumina B, Spadaro L, Trunfio G. Recent advances on wet air oxidation catalysts for treatment of industrial wastewaters. Inorganica Chimica Acta. 2015;**431**:101-109

[38] Bhargava SK, Tardio J, Prasad J, Föger K, Akolekar DB, Grocott SC. Wet oxidation and catalytic wet oxidation. Industrial & Engineering Chemistry Research. 2006;**45**(4):1221-1258

[39] M'Arimi MM, Mecha CA, Kiprop AK, Ramkat R. Recent trends in applications of advanced oxidation processes (AOPs) in bioenergy production. Renewable and Sustainable Energy Reviews. 2020;**121**:109669

[40] Holm-Nielsen JB, Al Seadi T, Oleskowicz-Popiel P. The future of anaerobic digestion and biogas utilization. Bioresource Technology. 2009;**100**(22):5478-5484

[41] Leng L, Yang L, Chen J, Leng S, Li H, Li H, et al. A review on pyrolysis of protein-rich biomass: Nitrogen transformation. Bioresource Technology. 2020;**315**:123801

[42] Qiu L, Deng YF, Wang F, Davaritouchaee M, Yao YQ. A review on biochar-mediated anaerobic digestion with enhanced methane recovery. Renewable and Sustainable Energy Reviews. 2019;**115**:109373

[43] Kamali M, Sweygers N, Al-Salem S, Appels L, Aminabhavi TM, Dewil R. Biochar for soil applications-sustainability aspects, challenges and

future prospects. Chemical Engineering Journal. 2022;**428**:131189

[44] Divya D, Gopinath LR, Merlin Christy P. A review on current aspects and diverse prospects for enhancing biogas production in sustainable means. Renewable and Sustainable Energy Reviews. 2015;**42**:690-699

[45] Mood H, Sohrab MR, Pelaez-Samaniego, Garcia-Perez M. Perspectives of engineered biochar for environmental applications: A review. Energy & Fuels. 2022;**36**(15):7940-7986

[46] Fang B, Qiao B, Zhao M, Jialei Guo Y, Wang YY, Chen H, et al. The accumulation of per-and polyfluoroalkyl substances (PFAS) at the air– water interface: A systematic review. Journal of Hazardous Materials. 2025;**138728**

[47] Ramos P, Schmidt MP, Xuan R, Ashworth DJ. Biochar selection for removal of perfluoroalkyl substances from reclaimed water for agricultural irrigation. Biochar. 2025;7(1):56

[48] Gao W, Lin Z, Chen H, Yan S, Huang Y, Xun H, et al. A review on N-doped biochar for enhanced water treatment and emerging applications. Fuel Processing Technology. 2022;**237**:107468

[49] Wang J, Wang S. Preparation, modification and environmental application of biochar: A review. Journal of Cleaner Production. 2019;**227**:1002-1022

Chapter 3

Recent Progress in the Influence of B- and N-Doped Primary and Secondary Hydrochar-Based Electrodes from Renewable Raw Materials on the Capacitive Performance for Energy Storage Applications

Jan Willem Straten

Abstract

A comparative analysis was conducted to elucidate the roles of primary char (PC) and secondary char (SC) formation in modulating the effects of B- and N-doping on the physical-chemical characteristics of the resulting electrodes for energy storage applications. For the synthesis of biobased N-doped hydrochar-based electrodes, sucrose (Suc), *Miscanthus* (Mis) and potato peels (PPs) were utilized as carbon sources, whereas urea and pyrrole (Pyr) were chosen as N precursors; boron trioxide (B_2O_3) was selected as B precursor. *In situ* doping *via* hydrothermal carbonization (HTC) was used, followed by pyrolysis, obtaining N-doped hydrochars (N-HCs). During HTC, the feedstocks undergo a multitude of complex decomposition mechanisms leading to the formation of N-HCs, comprising different ratios of PCs and SCs. Subsequent pyrolysis and chemical $ZnCl_2$ activation, produced N-doped pyrolyzed hydrochars (N-PHCs) with 5.6 wt % N from Suc and 4.8 wt % N from Mis applying urea. The N-PHCs from Suc and Mis employing Pyr, 3.1 and 4.2 wt % N, respectively were achieved. PPs-based syntheses focused mainly on activated carbons (ACs), B-doped AC yielded 0.2 wt% B, whereas co-doped B,N-AC reached 5.7 wt% N with 0.1 wt% B. Electrochemical testing revealed that specific capacitance increased, detecting 47.0 $F{\cdot}g^{-1}$ at 5 $mV{\cdot}s^{-1}$ for N-PHCs based on Suc and 3.5 $F{\cdot}g^{-1}$ at 100 $mV{\cdot}s^{-1}$ for those from Mis using urea. In case of PPs, the highest specific capacitances were recorded for B,N-co-doped AC, determining values of 51.7 $F{\cdot}g^{-1}$ at 100 $mV{\cdot}s^{-1}$ and 71.9 $F{\cdot}g^{-1}$ at 5 $mV{\cdot}s^{-1}$.

Keywords: hydrothermal synthesis, boron and nitrogen doping, primary and secondary chars, pyrolysis, energy storage

1. Introduction

In response to the growing global demand for sustainable and efficient energy storage devices, carbon-based materials derived from renewable raw materials have attracted close attention. When properly processed, these carbon materials can serve as high-performance electrodes for energy storage systems such as supercapacitors and other electrochemical devices. The turning away from fossil fuels, like coal, towards biobased feedstocks, including lignocellulosic biomass and carbohydrates, not only fosters the recycling economy but also lowers the environmental impacts related to conventional synthesis routes.

A promising environmentally friendly and cheap technique for transforming biomass into carbonaceous materials is hydrothermal carbonization (HTC). HTC occurs in aqueous medium at mild temperatures of usually 190 and 240°C under self-generated pressure, commonly known as subcritical water conditions. HTC produces hydrochar (HC) that resembles lignite in terms of structure and properties. The emergence and properties of HC through HTC depends on the nature of the biomass and the reaction conditions. Thus, the selection of the renewable resource is essential, as the chemical structure and composition are pivotal for material conversion. Depending on the structural complexity and decomposition behavior of the starting material, the HC is made up of solid char phases, which means either a mixture of primary char (PC) and secondary char (SC) or solely SC. Morphologically, in case of a blend of PC and SC, spherical particles of the SC accumulate on the PC surface. Generally, SC shows a higher C content and calorific value.

Due to their amorphous and chemically unstable nature, HCs often undergo a mild pyrolysis step or pyrolysis associated with chemical activation, generally performed between 500 and 900°C, in order to obtain thermally and chemically stable pyrolyzed hydrochars (PHCs) or, in conjunction with chemical activation, activated carbons (ACs) suitable for energy-related applications [1, 2]. By means of further modification, optimization, and functional tailoring, these biobased carbon materials can be processed into electrodes for use in electrochemical devices [3–5]. Notably, the incorporation of heteroatoms (*in situ* doping), e.g. nitrogen (N) or boron (B), has been proven to yield favorable material properties and structure–property relationships – for instance, improved surface functionalities or EC – for electrochemical applications [6]. Heteroatom-doped PHCs exhibit beneficial synergistic effects (e.g. electron donation or charge transport).

A large number of N-single-doped or B,N-co-doped carbon-based electrode materials for energy storage applications derived from earth-abundant renewable resources have already been investigated, e.g. shiitake mushroom or bean shell. They all demonstrated promising electrochemical performance for supercapacitor applications.

This review is based on the findings of three recently published research articles. It focuses on a comparative analysis of carbonaceous electrode materials synthesized from biogenic starting materials, namely *Miscanthus* (Mis), sucrose (Suc) and potato peels (PPs), emphasizing the influence of B- and N-doping. Mis and Suc were used in combination with urea and pyrrole (Pyr), whereas the B,N-co-doping strategy was performed with PPs in connection with boron trioxide (B_2O_3) and urea [7–9].

The first study implies the utilization of Mis and Suc in the context of *in situ* doping with the N-precursors urea and Pyr. An identical synthetic pathway is applied. The HTC and subsequent pyrolysis are carried out at the same reaction conditions, differing only in the N-dopant. The as-obtained N-doped PHCs (N-PHCs) were examined for their physical-chemical properties and their electrochemical performance. It aims to compare the material properties of N-PHCs, which contain various ratios of N-doped PC (N-PC) and N-doped SC (N-SC). Urea and Pyr are N-dopants typically applied to enhance supercapacitive behavior. These electrodes were investigated for their potential energy storage applications (**Figure 1**).

The second study deals with PPs. This synthesis route is different from the first study. Herein, the *in situ* doping involves B_2O_3 serving as the B-, and urea acting as the N-precursor. Urea and B_2O_3 are common precursors used to increase capacitive characteristics. Subsequent chemical activation with $ZnCl_2$ results in B-, and N-single and co-doped AC electrodes. This study sheds light on how the *in situ* doping followed by chemical activation with $ZnCl_2$ influences the physical-chemical features of the AC electrode materials for energy storage purposes (**Figure 2**).

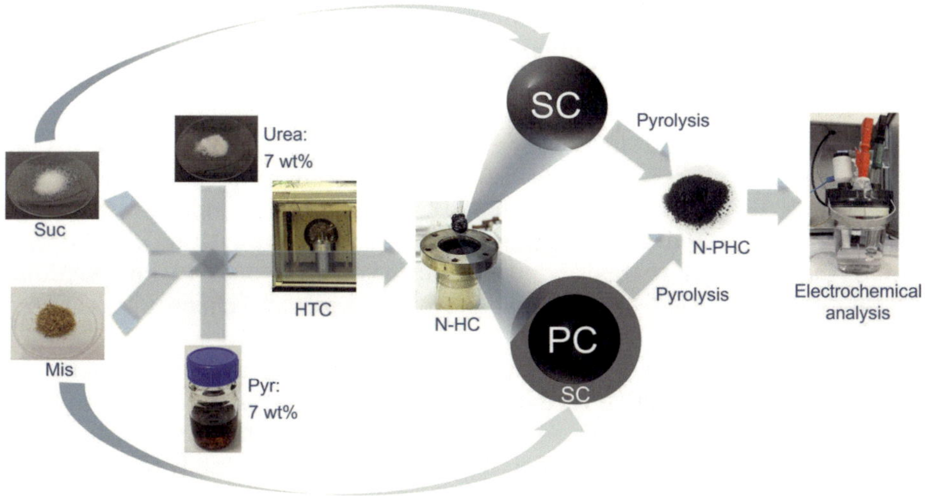

Figure 1.
Via HTC Suc and Mis were each added with urea and Pyr obtaining N-doped HCs (N-HCs) with PC and SC phases. Subsequent pyrolysis yields in N-PHCs for application in energy storage.

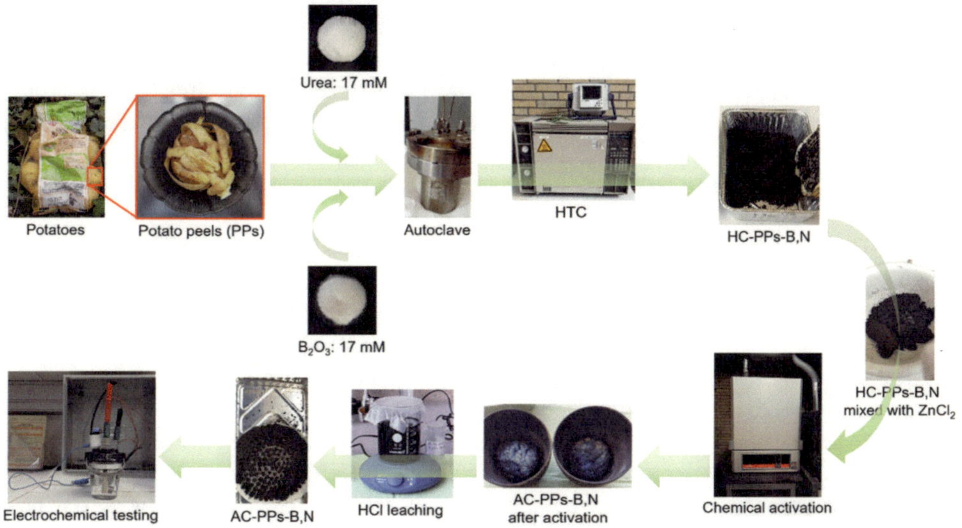

Figure 2.
Synthesis route of B,N-co-doped AC electrodes stemming from PPs.

The following sections discuss synthesis approaches, the role of PC and SC phases, the influence of distinct dopants on material properties, and the far-reaching consequences for sustainable energy storage.

2. The role of heteroatom-doped PC and SC char phases for energy storage applications

2.1 Synthesis strategies for biobased carbon electrodes

The fabrication of biobased carbonaceous electrode materials usually includes two steps. These steps involve HTC and subsequent pyrolysis, whereby pyrolysis and activation are typically combined. HTC is a thermochemical conversion process conducted in aqueous medium at subcritical conditions (190–240°C), transforming organic (waste) biomass into carbon-rich HC. HTC tailors morphology, chemical composition, structure and surface functionality of the HC. Therefore, the HTC is a suitable "green" synthesis technique offering an ideal environment for the *in situ* doping method. Briefly, the hydrothermal reactions imply decomposition reactions, namely hydrolysis, dehydration, fragmentation (e.g., decarboxylation), and polymerization reactions, such as (poly)condensation, as well as aromatization reactions. Depending on the type of feedstock, two morphologically and chemically distinct HC types emerge during HTC, PC and SC.

PC is produced from solid-solid carbonization reactions from the original structure of the raw material. It is characteristic of lignocellulosic biomasses like Mis or PPs. SC is formed from the aqueous phase through (re)polymerization and condensation reactions at the solid-liquid interface. This is especially prominent in carbohydrate-derived resources such as Suc.

After HTC, the HCs are either pyrolyzed or chemically activated (e.g. $ZnCl_2$) at higher temperatures (500–900°C). These conditions are crucial to adjust the desired porosity and specific surface area (SSA) of the electrode needed for energy storage. It is also employed to boost thermal and chemical stability, EC, and graphitization degree. These are significant factors to ameliorate electrochemical performance.

Moreover, the *in situ* doping is a decisive synthesis strategy to integrate heteroatoms into the carbon network. These heteroatoms provide active sites for desirable electrical properties, improved charge storage, and overall capacitance performance.

2.2 PC and SC formation from biogenic resources

Since the structural and chemical nature of PC and SC phases is determined by the type of feedstock, the differentiation is paramount for getting an idea about the doping behavior and the physicochemical features of the carbon materials.

Suc is a disaccharide and the hydrothermal reaction takes place at 190°C. The degradation starts with the hydrolysis of the α,β-1,2-glycosidic bond. The cleavage of the disaccharide results in the monosaccharides glucose and fructose. Via isomerization of glucose, fructose is formed. Fructose decomposes through dehydration reactions to the intermediate 5-hydroxymethylfurfural (HMF). HMF further decomposes *via* fragmentation reactions into organic acids (e.g. levulinic, acetic or formic acid). The organic acids and the byproducts (e.g. aldehydes, ketones) go through cyclization or (poly)condensation reactions in which cross-linking occur that lead to polyfuranic chains. These furanic structures react to aromatic-like carbon frameworks, aliphatic-based organic substances, gases and ultimately to solid polymeric HC. Suc is a carbohydrate, which means that SC predominates in HC, with PC being formed only to a lesser extent. In the liquid phase, (re)polymerization and condensation reactions prevail. Concurrent parallel reactions to volatile side products and the formation of organic byproducts dissolved in the liquid supernatant are formed to a greater or lesser extent depending on the synthesis route. Besides, numerous side reactions yield undesirable soluble (polymeric) organic byproducts known as humins.

Mis and PPs are lignocellulosic biomasses. Mis consists of cellulose (44–45%), hemicellulose (29–30%), and lignin (20–22%). PPs comprise cellulose (32%), hemicellulose (10%), starch (20%), lignin (20%), protein (18%), and lipids (1%).

During HTC, these main components influence the decomposition behavior. The emergence of HC encompasses two distinct pathways, the solved-intermediates-pathway as well as the solid-solid-conversion-pathway. The solved-intermediates-pathway is dictated by the polysaccharides cellulose, hemicellulose, and starch. Through the solid-solid-conversion-pathway, lignin is dominant. Hence, the HC formed from Mis and PPs are composed of PC and SC phases. The formation of HC from cellulose starts at temperatures above 220°C. The decomposition process initiates with the hydrolysis of the β-1,4-glycosidic bond into glucose. Afterwards, the degradation reactions are postulated to follow those of glucose. Compared to the above-described conversion mechanism of Suc *via* solved intermediates, there is an essential difference in the emergence of HC. The bulk of cellulose is directly transformed into an aromatic HC, which resembles the pyrolysis process.

2.3 Influence of N- and B-doping: Single and co-doping

Urea, Pyr and B_2O_3 are characteristic dopants used to improve physical-chemical material properties, such as the capacitive behavior.

Nitrogen is an n-type dopant and acts as an electron donor. Incorporated into the carbon skeleton, N modifies the surface chemical reactivity by enhancing wettability that is advantageous for both EC and capacitive performance. In contrast, B functions as a p-type dopant and is an electron acceptor. The presence of B raises the graphitization ratio in non-graphitizable carbons, thereby substantially increasing the EC. In addition, B contributes to elevated capacitive performance. The introduction of N and B leads to preferable synergistic effects. Co-doping (dual doping) provides a unique electron distribution and structural distortions that may enhance the electrochemical performance. B,N-co-doped carbon electrodes indicate pronounced defects within the carbon lattice and elevate structurally complex active sites. These active sites give superior chemical properties compared to their single-doped counterparts, implying higher reactivity, better catalytic, chemical, and electrochemical activity, ameliorated wettability for more effective electrode-electrolyte interaction, increased electronic properties, improved EC, enhanced energy density, reduced ohmic as well as charge transfer resistance, and excellent cyclic stability resulting in extended operational lifetime.

2.3.1 Chemical composition of HCs from Suc and Mis: Urea and Pyr as N-precursors

To investigate whether the *in situ* doping method with the N precursor urea succeeded, the elemental composition of all HCs was analyzed (**Figure 3**).

With a mass ratio of 7% of the N precursor urea, the N content in HCs increases to 6 wt% (**Figure 3**). The highest N content of 6 wt% is reached with Suc as feedstock.

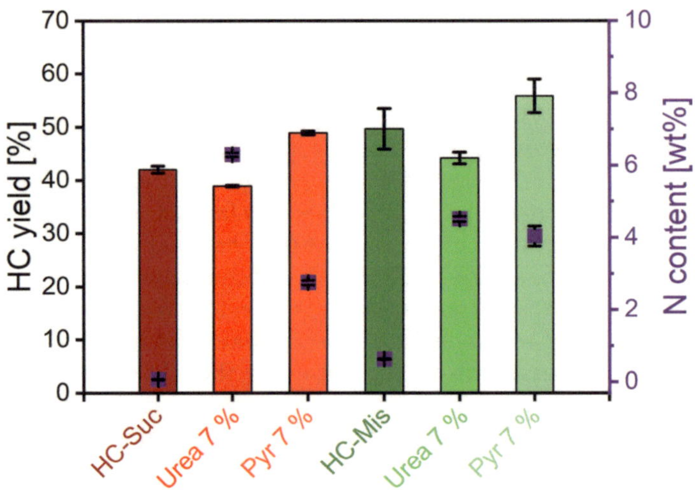

Figure 3.
HC yield (columns) and N content (squares) after HTC of Suc (red) and Mis (green) with the corresponding N precursors urea and Pyr.

N contents of up to 4 wt% are achieved using Mis as biogenic resource. If Pyr is used as N precursor with a mass ratio of 7%, the N content grows to 3 wt% for the HC stemming from Suc, whereas highest N content of 4 wt% is obtained utilizing Mis as starting material.

As the N content raises, the O content decreases in the cases of Suc and Mis applying both urea and Pyr. With respect to Mis, the C content drops slightly to (61.8 ± 0.2)% with growing urea content. If Suc is used as starting material, the increase in the C content is only marginal. In all cases, a decline in HC yield was achieved with the highest urea input of 7%. Conversely, if Pyr is used, the C content increases to (69.8 ± 0.1)% and (65.2 ± 0.8)% for HC from Suc and for HC based on Mis, respectively. Furthermore, a rise of HC yields was reached with highest Pyr addition of 7%. For the HC originating from Suc with 7% Pyr (HC-Suc-Pyr-7%), the HC yield is (48.9 ± 0.4)%, whereas for the HC from Mis with 7% Pyr (HC-Mis-Pyr-7%), highest HC yield of (55.8 ± 3.2)% was achieved.

Without the addition of N precursors, the HC yields are (42.0 ± 0.7)% and (49.7 ± 3.8)% for HC originating from Suc (HC-Suc) and for HC derived from Mis (HC-Mis), respectively. In **Figure 3**, the HC yield of HC-Suc is lower in comparison with HC-Mis. This can be accounted for by the various complexities of the decomposition mechanism during HTC. The degradation steps during the decomposition process follow a more or less complicated pathway depending on the choice of Suc or Mis as the substrate. Likely, parallel reactions to volatile side products and side reactions to soluble (polymeric) organic byproducts (e.g. humins) in the aqueous phase might be the reason for the lower HC yield of HC-Suc. However, a very high C content of (68.3 ± 0.1)% can be observed by means of elemental analysis. The high values of C are attributable to the fact that mainly O-functions are decomposed, which is why many O containing compounds are identified among the decomposition products. Yet, the degradation products of Suc contribute to the emergence of the carbon network to a greater extent. This can be a hint of a more pronounced internal condensation level and cross-linking of the carbon framework of HC-Suc.

Unlike HC-Suc, the HC yield of HC-Mis is lower, which is attributed to the emission of gaseous substances or to dissolved (polymeric) organic byproducts, in the same way as with Suc. Nonetheless, the decomposition products of Mis contribute to the formation of the carbonaceous framework to a lesser extent. The elemental analysis of HC-Mis also shows a lower C content (63.5 ± 0.01)%, which suggests a less pronounced internal condensation level and cross-linking of the carbon scaffold of HC-Mis.

Under subcritical water conditions, urea is decomposed into CO_2 and NH_3. Via Maillard reaction, NH_3 reacts with reducing sugar (e.g. Suc, decomposed hemicellulose). The Maillard reaction integrates N into the carbon scaffold and encompasses amination, imination (Schiff base), rearrangements, cyclization, degradation (e.g. Strecker degradation) reactions, or the transformation to aromatic compounds (e.g. pyrrole, pyrazine, furan, pyrone). The non-Maillard reactions are additional crucial reaction mechanisms to incorporate N into the carbon framework. They entail condensation (Mannich reaction), amidation (Schotten-Baumann reaction), (reductive) amination (Leuckart-Wallach, Eschweiler-Clarke reaction), or ring forming reactions (Paal-Knorr pyrrole and furane syntheses, Chichibabin pyridine synthesis or Staedel-Rügheimer pyrazine synthesis). Cyclization reactions also involve cycloadditions, mainly (hetero-)Diels-Alder reactions, which produce N-heterocycles. These diverse reaction mechanisms result in a multitude of O- and N-functional groups, linkages as well as structural motifs such as heterocycles. Since urea is easily hydrolyzed and is involved in a wide range of chemical reactions, this may be one reason why higher N contents are obtained.

The role of Pyr under subcritical conditions is not yet elucidated and understood. There are no mechanistic studies in the literature on Pyr in combination with hydrothermal synthesis, only in connection with polypyrrole (polyPyr) [10, 11]. In the first step, Pyr is chemically polymerized to polyPyr in the presence of an oxidizer ($FeCl_3$, ammonium persulfate or HCl) [11, 12]. In the second step, polyPyr is mixed with a carbon material (e.g. carbon cloth or graphene oxide) and hydrothermally treated to produce hybrid or composite materials [10, 11]. In a study, the mixture was added with Triton X-100® and hybrid or composite materials were synthesized *via* self-assembly hydrothermal treatment [11].

Consequently, it can probably be assumed that Pyr itself is only decomposed or hydrolyzed slowly, so that Pyr itself is incorporated into the polymeric carbonaceous network, and the Maillard reaction only plays a minor role in this degradation process. It is also reasonable to suggest that the N dopant might also be adsorbed on the carbon surface instead of being incorporated into the carbon skeleton.

In general, N-doping with urea and Pyr was successful as the N content rises to 6 wt% (**Figure 3**). In the cases for HC-Suc and HC-Mis, where the N content increases but the O content diminishes, this can be ascribed to the fact that O-functional groups are substituted by N-containing functions.

2.3.2 Chemical composition of HCs from PPs: Urea and B_2O_3 as N- and B-precursors

PPs are waste biomass consisting of a self-doped enrichment of N and O. As a result, PPs are promising starting materials for the transformation into high-performance, biobased carbonaceous electrodes for high added-value energy storage application.

The *in situ* doping method was examined with the B- and N-dopants urea and B_2O_3, respectively. The elemental analysis was employed to characterize all HCs and to figure out whether the *in situ* doping method was successful (**Figure 4**).

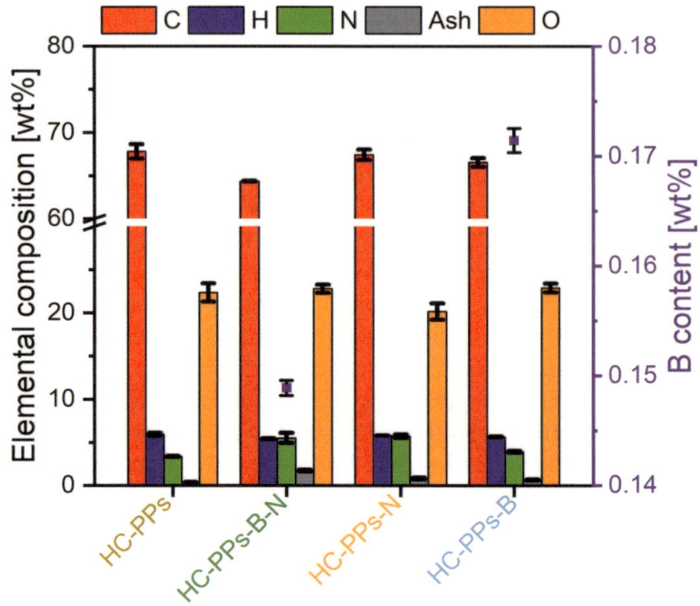

Figure 4.
Elemental composition, ash content (columns), and B-content (squares) of HCs from PPs with the B-dopant B_2O_3 and the N-dopant urea. The S content was not considered in the diagram (S is <0.1%).

The share of N of the HC stemming from PPs (HC-PPs) amounts to 3.4 wt%. With the addition of urea, the proportion of N increases. The HC comprising a blend of the precursors B_2O_3 and urea (HC-PPs-B-N), the share rises to 5.5 wt%. The maximum percentage of N of 5.7 wt% was obtained for the single-doped HC with urea (HC-PPs-N). With the addition of the B-dopant B_2O_3, there is no notable growth of the amount of B. For both HC-PPs-B-N and HC-PPs-B, the B proportion has a comparatively low content of 0.15 and 0.17 wt%, respectively. The share of C is comparatively high in all HCs fluctuating between 64 wt% and 68 wt%. The sample co-doped HC-PPs-B-N possesses the minimum percentage of C. The HC-PPs and HC-PPs-N samples display the maximum amount of C. For all HCs, the share of H is consistent at around 5 wt%, whereas the proportion of O is in a state of stagnation at about 20–23 wt%. Before and after hydrothermal synthesis, the ash content is noteworthy. PPs, dried in the cabinet, are characterized by an ash content of approximately 8.2 wt%. After the HTC, the ash content of all HCs decreased significantly to between 0.4 and 1.7 wt%. It is well-known that during hydrothermal synthesis, the ash content diminishes by leaching of inorganic substances to the aqueous medium.

The elemental analysis exhibits a fairly high content of C for all HC samples between 64 and 68 wt%. The reason for this is that O-containing structures are largely decomposed. This might be a hint for a more intense internal condensation level and cross-linking of the carbon framework of the HC samples.

The PPs and the precursors urea and B_2O_3 degrade under subcritical conditions. During the decomposition process and the associated transformation to HC, numerous reactions occur, particularly with the N-dopant urea that has already been described in-depth in the context of the decomposition process of Suc and Mis.

In fact, doping with N was successful, as the share of N rises. The sample HC-PPs illustrates an N proportion of 3.4 wt%. The sample single-doped HC-PPs-N indicates maximum N percentage of 5.7 wt%; co-doped HC-PPs-B-N has the second highest content of N with a value of 5.5 wt%.

Unlike N-doping, doping with B using B_2O_3 was unsuitable under subcritical conditions. The B content is extremely low at 0.15 and 0.17 wt% for the samples HC-PPs-B-N and HC-PPs-B, respectively. Accordingly, it can be stated that an extremely low amount of B is bonded in the carbon lattice of the HC. Nevertheless, investigations on reaction mechanisms with respect to B species under subcritical conditions are basically unknown. Some examinations assume that B structures bonded in carbon frameworks comprise substitutional B, boronic or borinic esters, and boronic acids. At temperatures below 800°C, inorganic B substances possess low doping efficiency because of their low reactivity. The role of the dopant B_2O_3 during hydrothermal synthesis remains unknown because of the lack of mechanistic investigations. It stands to reason that no appreciable degradation processes take place during hydrothermal synthesis. It can be postulated that once B_2O_3 is dissolved in water, metaboric acid and boric acid (H_3BO_3) are formed. Diols (e.g. hydroxycarboxylic acids) are among the degradation products of carbohydrates. In the presence of hydroxycarboxylic acids, boron acids (e.g. boric or boronic acids) can enter into esterification reactions to form boric ester (borate monoester) or boronic ester (boronate monoester).

2.3.3 Chemical composition of PHCs from Suc and Mis: Urea and Pyr as N-precursors

Pyrolysis at 900°C alters the chemical composition and properties, characterized by a substantial growth of the C share with a concurrent reduction in the proportion of H and O of the resulting PHCs. In contrast, the N content indicates only slight deviations from those of the HCs. The PHC yield and N content of the PHCs are presented in **Figure 5**.

The C content of the PHCs ranges from 81 to 94 wt%. The maximum C percentage of (94.0 ± 0.9) wt% is detected for PHC-Suc, whereas the lowest, (80.8 ± 0.2) wt%, is observed for PHC-Mis in the presence of urea at a mass ratio of 7 wt% (PHC-Mis-Urea 7%). The H content of all PHCs remains <1 wt%, while the N content lies

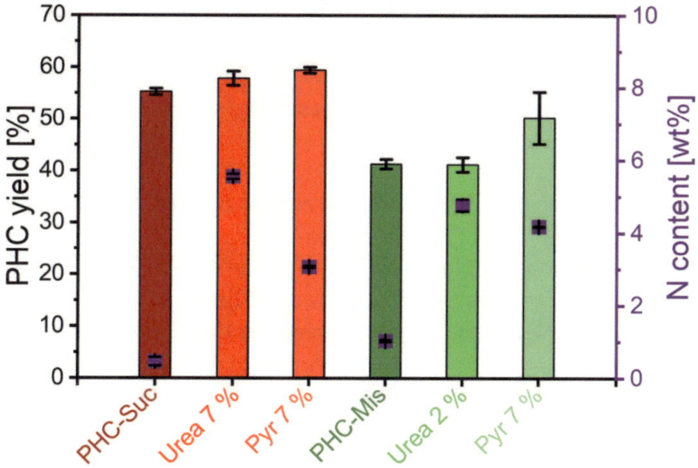

Figure 5.
PHC yield (columns) and N content (squares) after pyrolysis of Suc (red) and Mis (green) with the N-precursors urea and Pyr.

between 1 and 6 wt%. The maximum amount of N of (5.6 ± 0.07) wt% is achieved for PHC-Suc with 7 wt% urea (PHC-Suc-Urea 7%), and similarly, PHCs stemming from Mis with a mass ratio of 7 wt% urea (PHC-Mis-Urea 7%) indicate the highest N content of (4.8 ± 0.2) wt%. Generally speaking, the PHC yield based on both Suc and Mis increases with rising N content. However, the PHC yield of PHC-Mis-Urea 7% with a value of (41.2 ± 1.4)% is essentially comparable to that of its N-free counterpart (41.3 ± 0.9)%. Besides, the sample PHC-Suc-Urea 7% reaches a slightly higher PHC yield, (57.8 ± 1.4)%, relative to its N-free counterpart of (55.3 ± 0.6)%. When Pyr is used as a dopant, it is noticeable that the N content of the PHCs is slightly lower than when urea is utilized. On the contrary, the C content increases when Pyr is applied. No trend can be observed for the O and H contents. The highest PHC yields are recorded for the PHCs prepared with a mass ratio of 7 wt% Pyr. In particular, PHC-Suc with 7 wt% Pyr (PHC-Suc-Pyr 7%) achieves a PHC yield of (59.4 ± 0.6)%, whereas PHC-Mis with 7 wt% Pyr (PHC-Mis-Pyr 7%) yields a value of (50.2 ± 5.0)%. Among all PHCs, the maximum PHC yield is hence obtained for PHC-Suc-Pyr 7%. A trend is visible in the PHC yields. The higher the C content and the lower the share of O, with moderate N content, the higher the PHC yields. The trend is reversed for HC yields, in which the HC yields diminish with rising N content.

Potential structural entities present in the N-PHCs are thermostable connectivites such as quaternary N groups, pyrrole- or pyridine-like moieties. Unlike the PHCs originating from Suc, the PHC samples based on Mis exhibit lower average PHC yields, ranging from 41 to 50%. These PHC-Mis materials have average O contents stagnating between 10 and 14 wt% relative to the PHC-Suc materials possessing lower values of 5.2–5.6 wt%. The elevated O percentage may be ascribed to thermostable O functionalities, which are involved in chemical bonds and consequently contribute to cross-linking within the carbon scaffold, thereby imparting particular stability to the polymeric carbon framework. Likely structural motifs include phenolic, ether, and carbonyl groups, which are embedded as linkers within the polymer-like carbonaceous matrix. The sharp rise in C content, accompanied by a simultaneous diminution in O and H, can be related to the release of volatile species (H_2O, CO_2, CH_4) during pyrolysis; for instance, through cyclization reactions or *via* the annulation of (hetero)cyclic compounds into polycyclic aromatic ring systems. With regard to the PHC yields, it may be inferred that the N-SC in the PHC-Suc samples exerts a stabilizing effect on the carbon scaffold, resulting in higher PHC yields. In contrast to this, a mixture of N-PC and N-SC in the N-PHC-Mis materials does not seem to give any particular stability to the carbon skeleton, since the PHC yields remain essentially constant.

2.3.4 Chemical composition of ACs from PPs: Urea and B_2O_3 as N- and B-precursors

Following chemical activation of all HC samples with $ZnCl_2$ at 900°C, noticeable changes in the elemental composition can be seen (**Figure 6**).

The activation process leads to a slight increase in the C proportion, varying between 71 and 74 wt%, accompanied by a sharp decrease in the amount of H. On the other hand, the percentage of N, B, and O remains comparatively constant.

The sample AC-PPs-B-N unveils the maximum C content, yielding 74.1 wt%, the minimum value of 70.7 wt% is found for pristine AC-PPs. The N-content ranges from 3.1 to 5.7 wt%, with the maximum share of proportion of 5.7 wt% measured for AC-PPs-B-N and the minimum proportion of 3.1 wt% for AC-PPs-B. O contents remain at a stable level between 18 and 26 wt%, with AC-PPs-B representing the

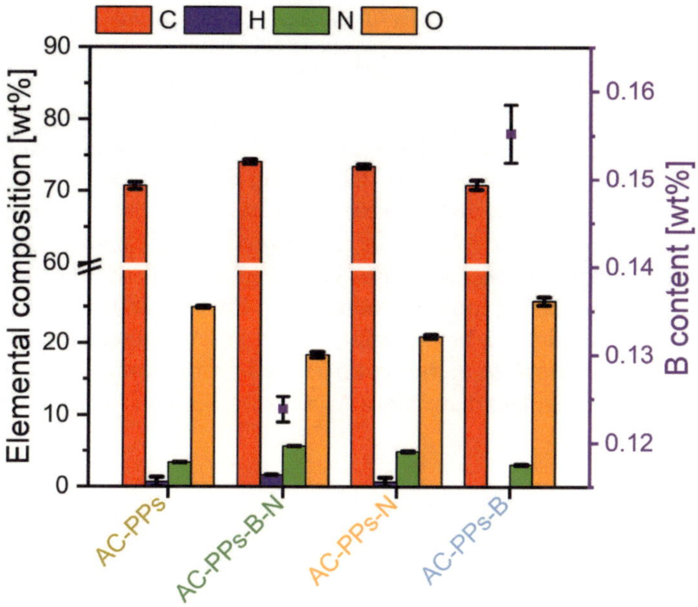

Figure 6.
Elemental composition (columns) and B content (squares) after chemical activation of B- and N-doped HC-PPs with $ZnCl_2$. The S content is not taken into consideration in the diagram (S is ≤0.3%).

highest amount of 25.8 wt% and AC-PPs-B-N possessing the lowest of 18.4 wt%. The values of the B percentage of AC-PPs-B-N and AC-PPs-B amount to 0.1 and 0.2 wt%, respectively. Of all determined chemical elements, H undergoes the most considerable decline upon pyrolysis. For AC-PPs and AC-PPs-B, H contents approach zero, with values of (0.6 ± 0.6) wt% and (0.7 ± 0.7) wt%, respectively. The sample AC-PPs-B does not contain any detectable H, while for AC-PPs-B-N a value of 1.6 wt% was measured.

The use of $ZnCl_2$ as a chemical activating agent, acting as a Lewis acid, is a dehydrant and facilitates carbonization reactions as well as promotes polycondensed aromatic ring reactions. Furthermore, $ZnCl_2$ reacts with water to generate an acidic medium that exerts a corrosive effect on the carbon matrix, in this way causing the formation of a porous, channel-like structure. Moreover, $ZnCl_2$ selectively emits volatile species such as H and O from the hydroaromatic structure, especially gaseous H_2O and H_2; hydrocarbons such as CO, CO_2, or O-containing organic components. This phenomenon is probably attributed to the significant decline in the share of H while simultaneously explaining the relatively constant proportion of O after chemical activation, since the emission of oxygenated volatiles is limited. Furthermore, the elevated O percentage can be deduced from the presence of thermostable O-structural units, for example phenolic, ether, and carbonyl groups that cross-link the carbon framework conferring structural stability. As a matter of fact, $ZnCl_2$ activation also promotes the production of benzoquinone and ether functional groups. The volatilization process creates reactive sites, in which cyclization reactions or annulation of ring structures into polycyclic aromatic compounds may take place. Besides, thermostable N- and B-components may account for the comparatively constant N and B content after chemical activation. N-structures are possibly existent in the form of quaternary-N, pyrrole, or pyridine, whereas B-structural links are probably present as substitutional B, boronic and borinic esters, or boronic acids, all of which remain stable at 900°C.

2.3.5 Electrochemical analysis of PHCs from Suc and Mis: Urea and Pyr as N-precursors

Cyclic voltammetry (CV) measurements were performed on PHC electrodes originating from Suc and Mis, as well as on their N-PHCs obtained with 7 wt% urea and 7 wt% Pyr, employing carbon black (CB) as the reference material. The PHC electrodes were assessed at two distinct scan rates, at a higher scan rate of 100 $mV{\cdot}s^{-1}$ and a lower scan rate of 5 $mV{\cdot}s^{-1}$. Representative anodic and cathodic half-cycles are identified within the applied potential interval of −0.2–1.2 V, with the anodic and cathodic half-cycles appearing more pronounced at lower scan rates (**Figure 7**).

For all carbon electrodes, higher scan rates yield in lower specific capacitances, whereas lower scan rates lead to higher specific capacitance values. At lower scan rates, the CV curves depict less distortion. As a result, higher specific capacitances are reached (**Figure 8**).

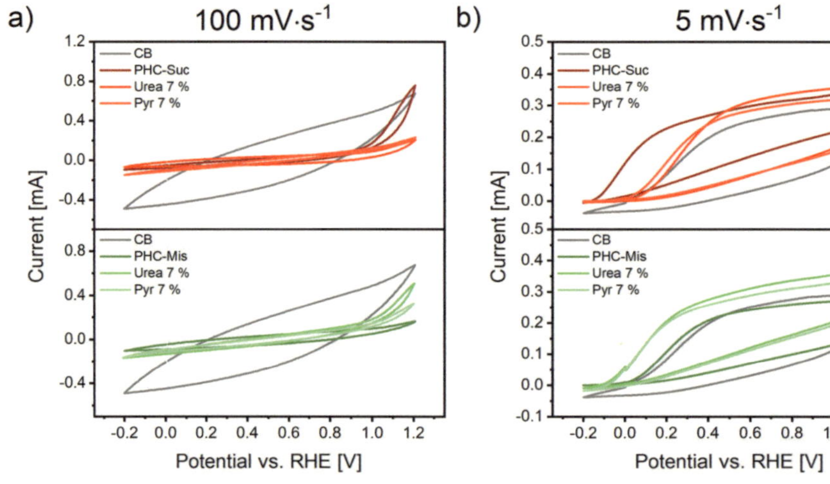

Figure 7.
Cyclic voltammograms of the PHC electrodes derived from Suc and Mis with the N precursors urea and Pyr at a scan rate of (a) 100 $mV{\cdot}s^{-1}$ and of (b) 5 $mV{\cdot}s^{-1}$ in 0.5 M H_2SO_4. CB is the reference material marked as gray line.

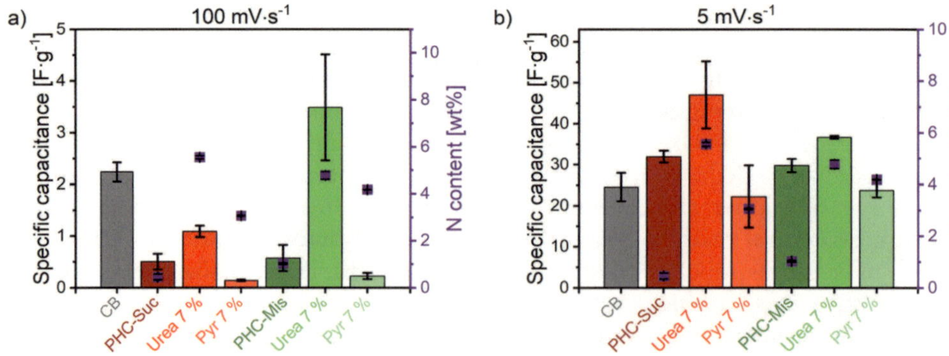

Figure 8.
Specific capacitances of the PHC electrodes (columns) from Suc (red) and Mis (green), as well as the N content (squares) with the N precursors urea and Pyr at a scan rate of (a) 100 $mV{\cdot}s^{-1}$ and of (b) 5 $mV{\cdot}s^{-1}$ in 0.5 M H_2SO_4. The reference material CB is denoted as gray column.

This phenomenon is due to the fact that more charge carriers (ions) penetrate into the pore network, which not only facilitates ion transport but also favors incorporation into the active carbon electrode. As a consequence, physical charge storage is enhanced through electrostatic interactions and physisorption, in conjunction with the development of an electric double layer (EDL) at the electrode/electrolyte interface (EEI). Previous reports note that both the shape of the CV curves and the specific capacitances of carbon electrodes deteriorate heavily with increasing scan rate. Considering that, the discussion section will focus on the results obtained at the lower scan rate of 5 mV·s^{-1}.

The maximum average specific capacitance was recorded for the electrode with the highest N content, namely PHC-Suc-Urea 7%, made up of N-SC, which achieved 47.0 F·g^{-1} at a scan rate of 5 mV·s^{-1}. Equally, PHC-Mis-Urea 7%, composed of a mixture of N-PC and N-SC, with a N content of 4.8 wt%, has the maximum specific capacitance of 3.5 F·g^{-1} at 100 mV·s^{-1} and the second-highest one of 36.7 F·g^{-1} at 5 mV·s^{-1}. In both cases, the values of the N-PHC electrodes exceed their undoped counterparts, obtaining superior specific capacitance at 100 mV·s^{-1} (1.1 ± 0.1 F·g^{-1} and 3.5 ± 1.0 F·g^{-1}), as well as at 5 mV·s^{-1} (47.0 ± 8.2 F·g^{-1} and 36.7 ± 0.3 F·g^{-1}).

The lowest specific capacitances are detected for N-PHC electrodes doped with Pyr. For PHC-Suc-Pyr 7%, the specific capacitance values are (0.1 ± 0.01) F·g^{-1} at 100 mV·s^{-1} and (22.2 ± 7.6) F·g^{-1} at 5 mV·s^{-1}. In the case of PHC-Mis-Pyr 7%, the corresponding values are (0.2 ± 0.06) F·g^{-1} at 100 mV·s^{-1} and (23.7 ± 1.6) F·g^{-1} at 5 mV·s^{-1}.

The cyclic voltammograms shown in **Figure 7** deviate from the representative rectangular profiles and instead demonstrate severely distorted, oval-shaped cycles with steep incline. This is owing to the existence of O- functionalities on the carbon surface, the low EC leading to high intrinsic resistance, and the concomitant slow electron-transfer rates. The resistive contributions also arise from the restricted ion mobility within the porous system. Although a reasonable amount of O bonded into the carbon lattice provides ameliorated surface wettability and pseudocapacitance, the influence of O-functions is notable, since their location at the edges of the carbon planes hinders electron transfer, thus raising electronic resistance. These impacts are evident across all examined PHC electrodes but are primarily marked for the PHC-Mis series, which reveal comparatively high O contents with values of 9.8–14.1 wt%. The elevated O contents not only evoke inferior EC values but also induce deficient specific capacitances (**Figure 8**).

Additionally, tiny humps are ascertained in all CV profiles within the potential range from 0.5–0.8 V for PHC-Suc; 0.7–1.0 V for PHC-Suc-Urea 7%; 0.6–0.9 V for PHC-Suc-Pyr 7%; 0.7–1.0 V for PHC-Mis; 0.6–0.9 V for PHC-Mis-Urea 7%, and 0.5–0.9 V for PHC-Mis-Pyr 7%, especially at a scan rate of 5 mV·s^{-1}. These observations are attributable to synergetic effects of electric double-layer capacitance (EDLC) and pseudocapacitance associated with N- or O-containing structural moieties. Nevertheless, the contribution from redox reaction peaks is minimal and barely discernible. The absence of apparent redox peaks is caused by faradaic charge transfer processes consisting of a vast number of reactions with innumerable onset potentials; there is no defined single reaction here. In view of the introduction of N-structural elements within the polymeric carbon network of the electrode, a variety of reactions between the N-active sites and protons from the acidic electrolyte have been proposed.

The electrodes N-PHC-Suc and N-PHC-Mis appear to be appropriate electrodes in terms of fairly high average specific capacitances with a value of 47.0 F·g^{-1} at 5 mV·s^{-1} and of 3.5 F·g^{-1} at 100 mV·s^{-1}, respectively. These electrodes represent the best-electrochemical performing material among all analyzed PHCs. The N-PHC-Suc electrode, containing SC, displays a sensible N content of 5.6 wt%, an EC of 129.1 $S{\cdot}m^{-1}$, and a superior average specific capacitance of 47 $F{\cdot}g^{-1}$ at 5 $mV{\cdot}s^{-1}$. The N-PHC-Mis electrode, composed of both PC and SC, features a N content of 4.8 wt% and an improved average specific capacitance of 3.5 $F{\cdot}g^{-1}$ at 100 $mV{\cdot}s^{-1}$, even though its EC is lower than that of the undoped counterpart. In this context, the predominance of SC in the N-PHC-Suc electrode seems to be beneficial for physical-chemical properties, as it leads not only to an increase in EC but also to specific capacitance at a lower scan rate. Similarly, SC is also present in the N-PHC-Mis electrode, albeit to a slightly lesser extent. Here, the same phenomenon can be observed. SC appears to be favorable for the physical-chemical properties, particularly the improvement in specific capacitance at a higher scan rate. Presumably, the presence of PC appears to have a negative effect on the EC, since the EC value is lower.

Besides, as determined, the carbon materials contain N and O. N-PHC-Suc illustrates a percentage of 5.6 wt% N and 5.2 wt% O, whereas N-PHC-Mis unveils a share of 4.8 wt% N and 14.1 wt% O. These heteroatoms are critical for capacitive effects, as basic surface N- and O-functions actively contribute to pseudocapacitive behavior through Faradaic redox reactions in acidic electrolyte. N-structural units entailing pyridinic N and quaternary N work as active sites to enhance electrochemical performance. Pyridinic N at the edge sites of the carbon lattice and quaternary N ameliorate electron donor properties by providing available electron pairs. These N-derived structures integrated into the carbon framework change the surface polarity that boosts wettability at the EEI and gives rise to pseudocapacitive effects. Ions diffuse preferentially through the pore system and enable beneficial access to the electrode surface, arousing higher specific capacitance.

Except for the amount of N and O, a definitive explanation for the highest specific capacitances of N-PHC-Suc and N-PHC-Mis is complicated. This finding cannot be explained by the results from the average pore size or total pore volume. N-PHC-Suc has an average pore size of 10.99 nm and a total pore volume of $1.02{\cdot}10^{-3}$ $cm^3{\cdot}g^{-1}$, whereas N-PHC-Mis possesses similar values. The average pore size accounts for 5.29 nm and the total pore volume is $1.18{\cdot}10^{-2}$ $cm^3{\cdot}g^{-1}$. N-PHC-Suc shows an overall negligible SSA of 0.19 and 4.48 $m^2{\cdot}g^{-1}$ for N-PHC-Mis accompanied by a relatively large average pore size and a low overall porosity. It is obvious that the comparatively large pore size favors diffusion of charge carriers to the carbon surface. Therefore, redox reactions are preferred, which eventually give superior specific capacitance. Nonetheless, an important reason for the inferior specific capacitance values is the generally very low SSA of all PHCs analyzed by BET measurements.

N-doping with Pyr did not provide any evident advantage in electrochemical performance, despite the paradoxically larger SSAs compared to the undoped counterparts. The PHC-Suc-Pyr 7% sample, made up of N-SC, reveals the highest SSA of 29.39 $m^2{\cdot}g^{-1}$, followed by PHC-Mis-Pyr 7%, consisting of a mixture of N-PC and N-SC, with an SSA of 13.63 $m^2{\cdot}g^{-1}$. The SSAs and total pore volumes of the N-PHC electrodes containing 7% Pyr surpass those of the N-PHCs prepared with 7% urea, while only the average pore sizes being smaller in comparison.

It should be noted that Pyr, despite the N-PHC electrodes having a higher N content, does not imply any merits concerning physical-chemical characteristics. Pyr does not appear to be convenient as an N-precursor under subcritical water conditions, since it might only hydrolyze slowly, whereby Pyr itself is embedded into the carbon matrix rather than being decomposed via Maillard reaction. Indeed, it stands to reason that Pyr might be absorbed on the carbon surface instead of being bonded into the carbon scaffold.

Although pyrolysis at 900°C was carried out, it does not appear to have wielded any considerable influence on the SSA of the PHCs. Yet, the maximum reached specific capacitance for N-PHC-Suc with a value of 47 $F{\cdot}g^{-1}$ at lower scan rate and 3.5 $F{\cdot}g^{-1}$ for N-PHC-Mis at higher scan rate remain quite lower than those reported for activated biobased carbon electrodes under comparable conditions. For instance, carbon electrodes based on Suc display specific capacitances ranging from 165 to 316 $F{\cdot}g^{-1}$ and SSAs between 518 and 2102 $m^2{\cdot}g^{-1}$ in acidic electrolyte. Apart from that, biomass-derived electrodes from Mis feature a specific capacitance value of 111 $F{\cdot}g^{-1}$ and an SSA of 3024 $m^2{\cdot}g^{-1}$ in organic electrolyte.

The primary reason for the poor electrochemical performance of the PHC electrode materials is the absence of chemical or physical activation to enlarge the SSA, which is typically required for supercapacitor applications. A further contributing factor is the weak adhesion of the PHC electrodes to the aluminum foil coated with carbon black (CB@Al current collector). The carrier film itself presents challenges, as the inhomogeneous distribution of the coating may cause uneven concentrations of physisorbed charge carriers at the EEI, whereby strongly impairing the performance of the EDL [4, 5]. For Mis-derived carbons, the biomass ash content may also play a role in diminishing specific capacitance. Depending on the type of Mis, the ash content varies between 2.2 and 3.5 wt%, which are relatively low values [13]. Nonetheless, its presence might deteriorate the electrochemical performance. Apart from that, the use of H_2SO_4 as the electrolyte likely influenced the low specific capacitance values. In this study, H_2SO_4 was prepared at a concentration of 0.5 M, whereas the highest specific capacitances are generally obtained at the optimal ionic conductivity of 1 M H_2SO_4 [4, 5, 14, 15]. Attempts to conduct electrochemical analysis in KOH were unsuccessful due to corrosion of the CB@Al current collector.

2.3.6 Electrochemical analysis of ACs from PPs: Urea and B_2O_3 as N- and B-precursors

Pristine AC electrodes from PPs, together with B- and N-single-doped as well as B,N-co-doped AC electrodes, were assessed by CV employing AC-Peat as a reference material. Measurements were performed in 1 M H_2SO_4 within the potential window of −0.2-1.2 V at a higher scan rate of 100 $mV{\cdot}s^{-1}$ and lower scan rate of 5 $mV{\cdot}s^{-1}$ (**Figure 9**).

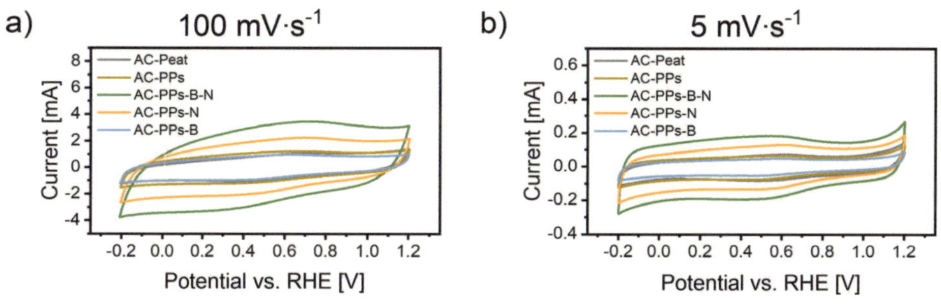

Figure 9.
CV profiles of the AC electrodes springing from PPs utilizing $ZnCl_2$ as chemical activator with the B- and N-dopants B_2O_3 and urea at a scan rates of (a) 100 $mV{\cdot}s^{-1}$ and of (b) 5 $mV{\cdot}s^{-1}$ in 1 M H_2SO_4. AC-Peat is the reference material.

The AC electrodes illustrate similar CV curves characterized by a quasi-rectangular shape within the applied potential window. Both anodic and cathodic half-cycles are each endowed with a broad hump incurred by redox reactions of B, N, and O present in functional groups acting as active sites. The respective redox peaks of the anodic and cathodic half-cycles can be assigned to the potentials compiled in **Table 1**.

As depicted in **Figure 9**, at a lower scan rate, the half-cycles seem to be more intense, since enhanced specific capacitance values can be found. In turn, at a higher scan rate, lower specific capacitances are identified (**Figure 10**). Increasing the scan rate modifies the CV shape. It is distorted with a slight inclination and, at the same time, the specific capacitances decrease. The CV profiles at 5 $mV{\cdot}s^{-1}$ present less distortion and the specific capacitance values are improved. This means a larger number of charge carriers (ions, electrons) can readily penetrate the pores of the active AC electrode. The average pore size of the ACs ranges from 0.95–1.38 nm, outperforming the dimensions of dissolved electrolyte ions SO_4^{2-} (5.33 Å) > $K^+ \approx H_3O^+$ (3.62–4.2 Å). Hence, the mobile charge carrier transport within the porous system is facilitated, allowing elevated physical charge storage through more extensive electrostatic interactions and physisorption, in this way, leading to a more pronounced EDL at the EEI.

Both at a high scan rate of 100 $mV{\cdot}s^{-1}$ and at a low scan rate of 5 $mV{\cdot}s^{-1}$, the AC-PPs-B-N electrode unveils the maximum average specific capacitances of 51.7

	100 $mV{\cdot}s^{-1}$		5 $mV{\cdot}s^{-1}$	
Sample	**Oxidation peak on the anodic half-cycle [V]**	**Reduction peak on the cathodic half-cycle [V]**	**Oxidation peak on the anodic half-cycle [V]**	**Reduction peak on the cathodic half-cycle [V]**
AC-Peat	0.73	0.41	0.61	0.51
AC-PPs	0.68	0.41	0.61	0.50
AC-PPs-B-N	0.71	0.23	0.54	0.48
AC-PPs-N	0.69	0.40	0.61	0.55
AC-PPs-B	0.64	0.41	0.60	0.49

Table 1.
Redox peak potentials of the AC electrode materials derived from PPs utilizing $ZnCl_2$ as chemical activator with the B- and N-dopants B_2O_3 and urea at a scan rate of 100 $mV{\cdot}s^{-1}$ and of 5 $mV{\cdot}s^{-1}$ in 1 M H_2SO_4. AC-Peat is the reference material [9].

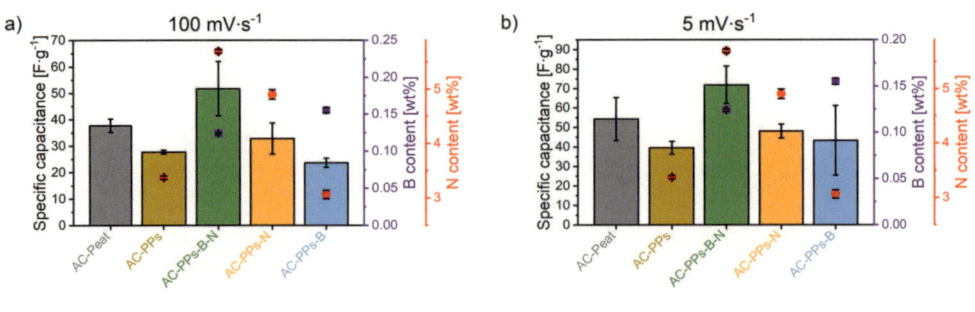

Figure 10.
Specific capacitance values of the AC electrodes stemming from PPs using $ZnCl_2$ as the chemical activator as well as the B- and N-dopants B_2O_3 and urea at a scan rate of (a) 100 $mV{\cdot}s^{-1}$ and of (b) 5 $mV{\cdot}s^{-1}$ in 1 M H_2SO_4. AC-Peat is the reference material.

and 71.9 $F \cdot g^{-1}$, respectively. Despite relatively large standard deviations, these values outperform the values of the reference material AC-Peat that delivered 37.7 ± 2.5 $F \cdot g^{-1}$ at 100 $mV \cdot s^{-1}$ and 54.3 ± 11.0 $F \cdot g^{-1}$ at 5 $mV \cdot s^{-1}$. The AC-PPs-N electrode provided the third-highest specific capacitance values of 32.8 and 48.1 $F \cdot g^{-1}$ at 100 and 5 $mV \cdot s^{-1}$, respectively. Pristine AC-PPs ranked fourth, with an average specific capacitance of 27.7 $F \cdot g^{-1}$ at 100 $mV \cdot s^{-1}$ and of 39.5 $F \cdot g^{-1}$ at 5 $mV \cdot s^{-1}$, the latter value being the lowest among all tested AC electrodes. The AC-PPs-B electrode produces the minimum specific capacitance with a value of 23.6 $F \cdot g^{-1}$ at 100 $mV \cdot s^{-1}$ and yields 43.3 $F \cdot g^{-1}$ at 5 $mV \cdot s^{-1}$, slightly surpassing that of pristine AC-PPs with a value of 39.5 $F \cdot g^{-1}$, although with a relatively high standard deviation. Taken as a whole, the results of B- and N-single-doped AC electrodes demonstrate enhanced specific capacitances compared to pristine AC-PPs. The electrode B,N-co-doped AC achieves the maximum specific capacitance at both scan rates, exceeding AC-Peat utilized as the reference material. The superior specific capacitance of the electrode AC-PPs-B-N can be explained by its maximum N proportion of 5.7 wt%, a minor B share of 0.1 wt%, a large SSA of 793.5 $m^2 \cdot g^{-1}$ with an average pore size of 0.95 nm, and an improved EC of 34 $S \cdot m^{-1}$.

In spite of the relatively small median pore size, the advantageous co-incorporation of the heteroatoms B and N alters the acid/base and electron donor/acceptor properties enabling charge carrier transport of the electrolyte ions within the porous AC electrode. The electrode B,N-co-doped AC changes the surface polarity of the carbon skeleton, and its hydrophilic character enhances electrode surface wettability in the electrolyte, simplifying ion access into the micropores. These synergetic effects accelerate ion kinetics, which increase the ion transfer efficiency and ease the electron-transfer ability. On account of their small ionic radii, the dissolved electrolyte ions can diffuse freely through the hierarchical pore systems that foster charge carrier transport within the carbonaceous electrode and across the EEI. Owing to the smaller ionic radii, the ion center approximates the AC electrode surface augmenting electrostatic interactions, which ultimately result in improved capacitive effects. These features are preferable for charge storage mechanisms at the EEI. **Figure 9** also depicts that O content substantially affects capacitive behavior. As is visible in **Figure 6**, the O proportion is between 18.4 and 25.8 wt%. In addition to B- and N-connectivities, O-structural moieties are implicated in pseudocapacitive performance through basic surface functions, which participate actively in reversible Faradaic redox reactions in acidic electrolyte. Among N-structural entities, pyridine, pyrrole, and quaternary-N are recognized as active sites. Pyridinic-N and pyrrolic-N, located at the edges of the carbon layer, provoke pseudocapacitive effects, while quaternary-N boosts electron transfer by donating its lone pair and thereby amending electron-donor properties. In terms of B-active sites, the precise Faradaic redox reaction mechanism is not yet completely understood. Nevertheless, the presence of B- and O-functionalities could contribute to reversible Faradaic redox reactions. Potential B-active sites imply substitutional B, boronic and borinic ester, or boronic acids, which, being electron-deficient, act as electron-withdrawing groups and increase the electron acceptor properties. Regarding O-active sites, benzoquinone structures and ether groups are likely candidates, as these functionalities are preferentially formed during chemical activation with $ZnCl_2$. In general, B-, N- and O-structural elements bonded into the carbon matrix alter the AC surface polarity and intensify wettability at the EEI triggering pseudocapacitance and enhancing rate performance. Therefore, charge carriers preferably migrate through the pore network to the electrode surface arousing an improved specific capacitance.

According to the pore textural parameters, AC-PPs-B-N accounts for an SSA of 793.5 $m^2 \cdot g^{-1}$, a median pore size of 0.95 nm as well as a total pore volume of 0.38 $cm^3 \cdot g^{-1}$, pointing to microporous system. Although these pore textural parameters are not as well-developed as those reported in the literature, the existence of B-, N-, and O-structural motifs bonded into the carbon framework plays a decisive factor in facilitating ion or electron diffusion through the pore system to the carbon surface. This advantageous diffusion behavior, coupled with favored redox reactions, ameliorates overall capacitive performance. Hence, the synergetic interactions of B-, N-, and O-structural units are of vital importance for the emergence of both EDLC and pseudocapacitance. It is important to state that a sensible O content can exert beneficial effects through O-functional groups, for example by amending surface wetting or pseudocapacitance. However, if a certain amount of O is surpassed, adverse impacts may result. The deviation from ideal rectangular CV profiles (**Figure 9**) is due to O-connectivities on the AC electrode surface, complicated by relatively low EC and high intrinsic resistance, which in turn arise from slow electron-transfer rates. Likewise, the resistance is introduced by the restricted mobility of charge carriers within the micropore network. Furthermore, O-structural elements induce high electron resistance and serve as barriers to electron transport. This phenomenon is observed across all AC electrodes, but is primarily pronounced for pristine AC-PPs, which represents the maximum amount of O with a value of 25.0 wt% (**Figure 6**). The elevated O content seems to be detrimental to the physical-chemical characteristics, since AC-PPs features the minimum EC of all characterized ACs and the minimum specific capacitance value at 5 $mV \cdot s^{-1}$. In the CV curves, little but discernible redox peaks are detected. The absence of well-defined peaks is on account of Faradaic charge transfer mechanisms encompassing a myriad of reactions with a multitude of onset potentials. The presence of N-structures within the carbon network, together with protons from the acidic electrolyte, can give rise to a great number of redox reactions rather than a single reaction.

The findings of the physical-chemical outcome were compared with literature studies on PPs-derived AC electrodes. Reported SSAs range from 103 to 1911 $m^2 \cdot g^{-1}$, with specific capacitances between 61 and 603 $F \cdot g^{-1}$ in acidic electrolytes; in neutral medium, the specific capacitances are 98–323 $F \cdot g^{-1}$.

In this study, the SSAs achieved with values of 627 to 825 $m^2 \cdot g^{-1}$ fall within the literature range. Nonetheless, the highest specific capacitances obtained with 51.7 $F \cdot g^{-1}$ at 100 $mV \cdot s^{-1}$ and 71.9 $F \cdot g^{-1}$ at 5 $mV \cdot s^{-1}$ do not correlate entirely with the values in the literature. These results highlight that the synthesis strategy for PPs-based AC electrodes still offers ample opportunities for optimization.

2.4 Scalability and reproducibility

The global transition to a circular and climate-neutral energy economy requires both performance-optimized materials and environmentally friendly synthesis strategies. In this context, the usage of biobased raw materials such as Mis, Suc, and PPs represents an attractive and sustainable option to fossil fuels (e.g. coal).

2.4.1 Biomass and dopant selection

Mis is a resistant, perennial C_4 grass with high biomass productivity, efficient CO_2 fixation, and low agricultural input requirements. Its robust lignocellulosic structure allows carbonization into structurally stable chars and offers long-term availability for scalable material production.

Suc is less lignified but a globally abundant carbohydrate and serves a model compound suitable for mechanistic studies, as well as for fine-tuning of SC formation. Suc is a food resource and thus less ideal for large-scale applications. However, its structural simplicity and high reactivity make it a valuable template for material development.

PPs are a major industrial waste stream, accounting for 6–10% of global potato processing volume. Their conversion into biobased carbon electrodes for energy storage technologies would be a key factor for high added-value applications or large-scale applications, avoiding landfill disposal or processing into fodder or fertilizer.

Urea is a cheap and harmless N-source widely used in agricultural and industrial applications. It is an appealing candidate for large-scale applications.

Pyr has some limitations with respect to toxicity, volatility, and lower incorporation efficiency, which would restrict large-scale applications.

B_2O_3 exhibits low doping efficiency under subcritical water conditions but may cause advantageous electronic effects at low concentrations. Nevertheless, large-scale applications are limited.

2.4.2 HTC and chemical activating agent

The hydrothermal synthesis employed across all studies is characterized by considerable ecological merits. HTC is processed at low temperatures (190–240°C), it is operated in aqueous medium without the need of organic solvents, and the *in situ* heteroatom doping is beneficial during char formation, thus avoiding post-treatment steps. Another advantage is the minimal gas emissions compared to conventional pyrolysis processes.

The utilization of $ZnCl_2$ for chemical activation raises concerns about corrosiveness and waste disposal. Yet, with convenient recycling strategies and wastewater treatment, the environmental impacts can be alleviated.

Overall, the combination of waste biomass (PPs), harmless dopants (urea) and scalable HTC processes offers a promising path to environmentally friendly biogenic electrode materials. The results underline that, when evaluating materials for energy storage applications of the future, sustainability must be taken into account alongside performance.

3. Conclusions

This presented study provides a comparative perspective on the synthesis, structural properties, and electrochemical performance of B-and N-single-doped as well as B,N-co-doped HC-based electrode materials derived from biogenic raw materials, including Suc, Mis and PPs. The investigation systematically assesses Suc, Mis and PPs, and their conversion into doped carbon materials via HTC, pyrolysis, and chemical activation. Briefly, the key findings are summarized below.

When doped with urea using a mass ratio of 7%, Suc forms an N-PHC that consists of SC with an N-content of 5.6 wt%. The N-PHC-Suc-Urea 7% electrode, comprising SC, unveils the highest EC of (129.1 ± 17.9) $S{\cdot}cm^{-1}$, and CV at a low scan rate of 5 $mV{\cdot}s^{-1}$ yields the maximum specific capacitance of (47.0 ± 8.2) $F{\cdot}g^{-1}$. These enhancements in EC and specific capacitance corroborate their potential as high-performance electrode materials.

Mis, doped with urea applying a mass ratio of 7%, produces a mixture of PC and SC with a N share of 4.8 wt%. The PHC-Mis-Urea 7% electrode, which is made up of a blend of PC and SC, achieves the highest specific capacitance at a high scan rate of 100 mV·s^{-1}, with a value of (47.0 ± 8.2) F·g^{-1}. The improvement in the specific capacitance confirms this electrode to be a promising candidate for advanced electrode material.

With respect to Pyr, after pyrolysis, PHC-Suc-Pyr 7% and PHC-Mis-Pyr 7% demonstrate a N percentage of 3.1 and 4.2 wt%, respectively. Pyr does not appear to be a suitable N-dopant, since it yields lower N content and reduced electrochemical performance. The specific capacitances drop to (0.1 ± 0.01) F·g^{-1} at 100 mV·s^{-1} and (22.2 ± 7.6) F·g^{-1} at 5 mV·s^{-1} for PHC-Suc-Pyr 7%. For PHC-Mis-Pyr 7%, at 100 mV·s^{-1} and 5 mV·s^{-1}, the specific capacitance is (0.2 ± 0.06) F·g^{-1} and (23.7 ± 1.6) F·g^{-1}, respectively.

In PPs-based electrodes, after activation with $ZnCl_2$, the highest N content is recorded for AC-PPs-B-N with a value of 5.7 wt% and for AC-PPs-N with 4.9 wt%. Despite the low doping efficiency of B_2O_3, with AC-PPs-B-N and AC-PPs-B reaches 0.1 and 0.2 wt%, respectively, both B- and N-doping affect the physical-chemical features. The AC-PPs-B-N electrode possesses the best electrochemical behavior, resulting in a specific capacitance of 51.7 F·g^{-1} at a higher scan rate of 100 mV·s^{-1} and 71.9 F·g^{-1} at a lower scan rate of 5 mV·s^{-1}. The outcome stresses that B,N-co-doping has positive impacts on the specific capacitance owing to synergistic effects.

In all, the HTC route enables sustainable char production using renewable and abundant feedstocks. Among the electrodes studied, urea-doped PHC made from Suc and B,N-co-doped AC based on PPs prove to be appealing candidates for energy storage applications and might pave the way for the production of next-generation sustainable electrodes.

Author details

Jan Willem Straten
Department of Conversion Technologies of Biobased Resources (440f), University of Hohenheim, Institute of Agricultural Engineering, Stuttgart, Germany

*Address all correspondence to: jan.straten@uni-hohenheim.de; jan.straten1@gmail.com

References

[1] Yoon Y, Lee M, Kim SK, Bae G, Song W, Myung S, et al. A strategy for synthesis of carbon nitride induced chemically doped 2D MXene for high-performance supercapacitor electrodes. Advanced Energy Materials. 2018;**8**:1703173. DOI: 10.1002/aenm.201703173

[2] Wang Z, Jin B, Peng J, Su W, Zhang K, Hu X, et al. Engineered polymeric carbon nitride additive for energy storage materials: A review. Advanced Functional Materials. 2021;**31**:2102300. DOI: 10.1002/adfm.202102300

[3] Sekhon SS, Park J-S. Biomass-derived N-doped porous carbon nanosheets for energy technologies. Chemical Engineering Journal. 2021;**425**:1-19. DOI: 10.1016/j.cej.2021.129017

[4] Alhnidi M-J, Straten JW, Nicolae SA, Hoffmann V, Titirici M-M, Kruse A. Thermal treatment versus hydrothermal carbonization: How to synthesize nitrogen-enriched carbon materials for energy storage applications? International Journal of Energy Research. 2022;**46**:1622-1636. DOI: 10.1002/er.7275

[5] Konnerth P, Jung D, Straten JW, Raffelt K, Kruse A. Metal oxide-doped activated carbons from bakery waste and coffee grounds for application in supercapacitors. Materials Science and Technology. 2021;**4**:69-80. DOI: 10.1016/j.mset.2020.12.008

[6] Titirici M-M, White RJ, Zhao L. Nitrogen-doped hydrothermal carbons. Green. 2012;**2**:25-40. DOI: 10.1515/green-2011-0023

[7] Straten JW, Alhnidi M-J, Mustaka E, Alchoumari G, Jung D, Kruse A. A comparative study on the influence of N-doped primary and secondary hydrochars on the electrochemical performance of biobased carbon electrodes for energy storage application. ChemCatChem. 2024;**16**:e202400142. DOI: 10.1002/cctc.202400142

[8] Straten JW, Alhnidi M-J, Mustaka E. Biobased N-doped primary and secondary char electrodes for energy storage application using pyrrole as the N precursor. Energy & Fuels. 2024;**38**:21661-21664. DOI: 10.1021/acs.energyfuels.4c03419

[9] Straten JW, Alhnidi M-J, Alchoumari G, Sangam K, Kruse A. B,N-doped activated carbon-based electrodes from potato peels for energy storage applications. ChemistryOpen. 2025;**14**:e202400527. DOI: 10.1002/open.202400527

[10] Arvas MB. Hydrothermal synthesis of polypyrrole/dye-functionalized carbon cloth electrode for wide potential window supercapacitor. Synthetic Metals. 2023;**293**:117275. DOI: 10.1016/j.synthmet.2022.117275

[11] Moyseowicz A, Pająk K, Gajewska K, Gryglewicz G. Synthesis of polypyrrole/reduced graphene oxide hybrids via hydrothermal treatment for energy storage applications. Materials. 2020;**13**:2273. DOI: 10.3390/ma13102273

[12] Tan Y, Ghandi K. Kinetics and mechanism of pyrrole chemical polymerization. Synthetic Metals. 2013;**175**:183-191. DOI: 10.1016/j.synthmet.2013.05.014

[13] Brosse N, Dufour A, Meng X, Sun Q, Ragauskas A. Miscanthus a fast-growing crop for biofuels and chemicals production. Biofuels, Bioproducts

and Biorefining. 2012;**6**:580-598. DOI: 10.1002/bbb.1353

[14] Rufford TE, Hulicova-Jurcakova D, Zhu Z, Lu GQ. Nanoporous carbon electrode from waste coffee beans for high performance supercapacitors. Electrochemistry Communications. 2008;**10**:1594-1597. DOI: 10.1016/j.elecom.2008.08.022

[15] Zhong C, Deng Y, Hu W, Qiao J, Zhang L, Zhang J. A review of electrolyte materials and compositions for electrochemical supercapacitors. Chemical Society Reviews. 2015;**44**:7484-7539. DOI: 10.1039/C5CS00303B

Chapter 4

Isotopic Tracers and Biochar in Soil Fertility and Plant Nutrition: Mechanistic Insights into Nutrient Cycling, Carbon Dynamics and Agroecosystem Functioning – A Review

Rodrigo Nogueira de Sousa, Giovani Rangel Malerba Alfonso and Kassio Ferreira Mendes

Abstract

Biochar has emerged as a multifunctional soil amendment with the potential to improve soil fertility, enhance nutrient use efficiency, mitigate contaminant mobility and contribute to long-term carbon sequestration. Produced through the pyrolysis of biomass under oxygen-limited conditions, biochar is characterized by a highly aromatic carbon structure, variable mineral composition and a chemically heterogeneous surface rich in oxygen-containing functional groups. These properties underpin its strong interactions with soil minerals, nutrients, microorganisms, and organic compounds. The mechanistic understanding of how biochar modifies biogeochemical processes in soil-plant systems remains incomplete. A major limitation in biochar research is the difficulty in disentangling direct physicochemical mechanisms, such as sorption and surface complexation, from indirect biological processes mediated by changes in microbial activity, nutrient cycling and rhizosphere dynamics. Conventional soil analyses largely rely on static measurements of nutrient pools and provide limited insight into fluxes, turnover rates and source-sink relationships. In this context, isotopic techniques employing stable and radioactive isotopes offer a powerful framework to resolve carbon, nutrient and contaminant dynamics in biochar-amended soils with high mechanistic resolution. This chapter synthesizes the principles, methodologies and applications of isotopic tracers in studies of soil fertility and plant nutrition under biochar amendment. Emphasis is placed on the use of ^{13}C and ^{14}C to elucidate carbon cycling and priming effects, ^{15}N to quantify nitrogen transformations and biological nitrogen fixation and radioisotopes such as ^{32}P, ^{33}P and ^{35}S to assess nutrient availability and turnover. Isotopic approaches are also discussed in the context of micronutrient dynamics and contaminant fate.

Keywords: radioactivity, priming effects, nutrients fluxes, rhizosphere, ^{14}C radioisotope

IntechOpen

1. Introduction

Biochar has gained global prominence as a soil amendment capable of modifying soil chemical, physical, and biological properties, with implications that extend from crop productivity to environmental quality and long-term carbon sequestration. Produced through the pyrolysis of biomass under limited oxygen availability, biochar is characterized by a highly condensed aromatic carbon structure, variable mineral content and a heterogeneous surface chemistry dominated by oxygen-containing functional groups [1, 2]. These properties collectively confer biochar a unique capacity to interact with soil constituents, nutrients, microorganisms and organic compounds [3]. Nevertheless, despite the extensive number of agronomic and environmental studies reporting positive or neutral effects of biochar application, the mechanistic basis underlying many of these responses remains incompletely resolved.

A central challenge in biochar research is the difficulty in disentangling direct physicochemical effects such as sorption, ion exchange and surface complexation from indirect biological effects mediated by shifts in microbial activity, nutrient cycling and rhizosphere processes. Conventional soil chemical analyses typically quantify changes in nutrient pools or extractable fractions [4] but are inherently limited in their ability to identify fluxes, transformation pathways and source-sink relationships. In this context, isotopic techniques, employing both stable and radioactive isotopes, represent a powerful and underutilized toolbox to resolve the dynamics of carbon and nutrients in biochar-amended systems with unparalleled resolution.

Isotopic tracers have a long tradition in soil fertility and plant nutrition research, enabling the quantification of nutrient availability, biological nitrogen fixation, mineralization, immobilization turnover, root uptake efficiency and contaminant fate [5]. When integrated with biochar research, isotopic approaches allow researchers to discriminate between native soil-derived elements and those originating from fertilizers, amendments or organic inputs, to quantify isotopically exchangeable pools, and to track the transformation and mobility of nutrients and agrochemicals over time. This chapter synthesizes the principles, methodologies and applications of isotopic tracers in the study of soil fertility and plant nutrition in biochar-amended systems, emphasizing mechanistic insights into carbon cycling, macronutrient and micronutrient dynamics and contaminant interactions.

2. Fundamental principles of isotopic tracers in soil: Plant systems

Isotopes are atoms of the same chemical element that differ in neutron number and, consequently, in atomic mass. This difference does not substantially alter chemical behavior but enables the discrimination of isotopic species during physical or nuclear detection [6]. Stable isotopes, such as ^{13}C, ^{15}N and ^{34}S, do not undergo radioactive decay and are quantified by measuring isotopic ratios relative to international standards using isotope ratio mass spectrometry (IRMS). Radioisotopes, including ^{14}C, ^{32}P, ^{33}P, ^{35}S and ^{65}Zn, possess unstable nuclei and emit ionizing radiation during decay, which can be detected by liquid scintillation counting (LSC) or gamma spectrometry, depending on the emission characteristics.

The power of isotopic tracers lies in their ability to follow the movement and transformation of an element through complex systems without significantly perturbing the system itself [7].

In soil fertility research, isotopes are used to quantify nutrient fluxes rather than static pools, providing direct measurements of availability, turnover rates and source contributions [8]. Two conceptual approaches dominate isotopic studies: enrichment techniques, in which a system is deliberately labeled with an isotope at levels above natural abundance, and natural abundance approaches, which exploit inherent isotopic differences among sources or processes.

Isotopic dilution constitutes one of the most important quantitative frameworks in soil fertility studies. In the case of phosphorus, the use of ^{32}P or ^{33}P allows the determination of the isotopically exchangeable P pool (E-value), reflecting the fraction of soil P that is in rapid equilibrium with soil solution and potentially available to plants [9]. Similarly, ^{15}N dilution methods enable the quantification of biological nitrogen fixation by comparing isotopic enrichment in fixing and non-fixing plants [10]. These approaches are particularly relevant in biochar-amended soils, where sorption processes, microbial immobilization and altered soil chemistry can substantially modify nutrient exchange dynamics.

3. Biochar properties and their isotopic implications

The intrinsic properties of biochar are central to understanding its interactions with isotopic tracers. One of the most relevant features is its carbon isotopic signature. Biochar produced from C_4 biomass, such as maize or sugarcane residues, exhibits a $\delta^{13}C$ value typically around 12‰, whereas soils dominated by C_3 vegetation generally present $\delta^{13}C$ values near 26 to 28‰ [11]. This contrast allows biochar-derived carbon to be traced in soil organic matter pools using natural abundance ^{13}C techniques, without the need for artificial labeling. Such an approach has been extensively applied to quantify the contribution of biochar carbon to total soil carbon stocks, microbial biomass and respired CO_2 [12].

Beyond isotopic signatures, biochar's surface chemistry exerts profound effects on nutrient behavior. Carboxyl, phenolic and lactonic functional groups, along with mineral phases inherited from the feedstock or formed during pyrolysis, create reactive surfaces capable of binding anions and cations [1]. These interactions influence the recovery and apparent availability of isotopically labeled nutrients. For example, phosphate labeled with ^{32}P may be rapidly sorbed onto biochar surfaces or biochar-associated metal oxides, reducing its immediate isotopic exchangeability. Conversely, biochar-induced increases in soil pH may reduce P fixation by Fe and Al oxides in highly weathered soils, increasing the isotopically exchangeable P pool [13].

Biochar also modifies soil microbial habitats by increasing porosity, altering moisture retention and supplying labile organic compounds, particularly in low-temperature biochars [14]. These changes often result in increased microbial biomass and activity, which can rapidly immobilize isotopically labeled nutrients such as ^{15}N or ^{33}P. Consequently, isotopic studies in biochar-amended soils must carefully consider microbial assimilation as a transient but critical sink influencing nutrient availability to plants.

4. Carbon cycling in biochar-amended soils traced by ^{13}C and ^{14}C

Carbon dynamics represent one of the most intensively studied aspects of biochar research, and isotopic techniques are indispensable for distinguishing biochar-derived carbon from native soil organic matter. Natural abundance ^{13}C has been widely used to partition soil carbon pools and CO_2 emissions following biochar application [15], enabling the quantification of biochar mineralization rates and the assessment of priming effects on native organic matter.

Radiocarbon (^{14}C) techniques provide even greater resolution. When biochar is artificially labeled with ^{14}C before soil application, it becomes possible to directly quantify its mineralization by measuring ^{14}C-CO_2 evolution over time [16]. Such studies have consistently demonstrated the remarkable stability of biochar carbon, with mineralization rates typically orders of magnitude lower than those of fresh plant residues. Moreover, ^{14}C partitioning approaches have revealed that biochar can either suppress or stimulate the decomposition of native soil organic matter, depending on biochar properties, soil type and environmental conditions.

Figure 1 illustrates how total CO_2 efflux can be quantitatively resolved into source-specific contributions, demonstrating that biochar addition does not simply alter the magnitude of soil respiration but fundamentally modifies the origin of mineralized

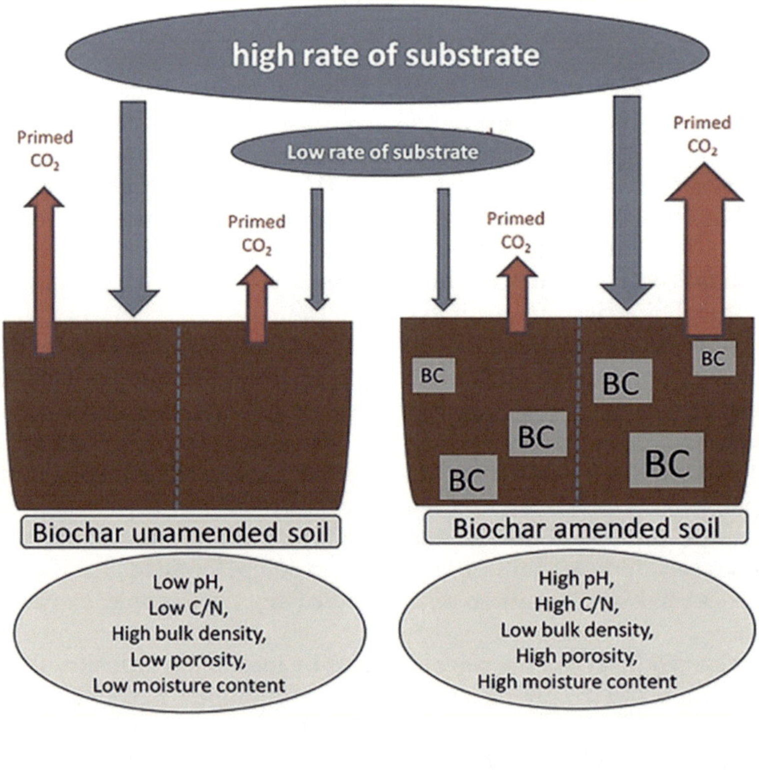

Figure 1.
Conceptual representation of carbon source partitioning in biochar-amended soils using a combined isotopic approach. The schematic illustrates the integration of ^{14}C radiolabeling and ^{13}C natural abundance to simultaneously quantify CO_2 derived from three distinct carbon sources: native soil organic carbon (C_3-derived), biochar carbon produced from C_4 biomass, and an easily degradable external substrate (^{14}C-labeled glucose). This approach allows the resolution of total soil CO_2 efflux into source-specific contributions, enabling mechanistic identification of biochar-induced priming effects and revealing how fresh carbon inputs modify both the magnitude and origin of mineralized carbon in biochar-enriched soil systems. Source: Adapted from Luo et al. [16].

DOI: http://dx.doi.org/10.5772/intechopen.1014824

carbon, particularly under conditions of fresh carbon inputs. By explicitly separating CO_2 derived from soil organic matter, biochar and added substrate, the schematic highlights the occurrence of biochar-induced priming effects and reveals that labile carbon availability can stimulate not only soil organic matter decomposition but also measurable losses of biochar carbon itself. This integrated isotopic approach overcomes the limitations of traditional two-source partitioning methods based solely on stable isotopes and provides a robust mechanistic framework to investigate carbon stabilization and destabilization processes in biochar-enriched soils, reinforcing the central role of isotopic tracers in advancing the understanding of soil carbon dynamics beyond bulk CO_2 flux measurements.

In plant-soil systems, isotopic approaches allow the discrimination of root-derived carbon from biochar-derived carbon within the rhizosphere. This distinction is essential for understanding how biochar influences rhizodeposition, microbial utilization of root exudates and the formation of stable organo-mineral associations. Accelerator mass spectrometry (AMS) has emerged as a powerful tool in this context, enabling ultra-sensitive detection of ^{14}C in complex matrices and facilitating long-term studies of carbon stabilization.

5. Nitrogen dynamics and ^{15}N tracing in biochar-amended systems

Nitrogen cycling is influenced by biochar through mechanisms that include ammonium adsorption, altered nitrification rates, changes in soil aeration and shifts in microbial community structure [17]. The use of ^{15}N-labeled fertilizers provides a direct means to quantify these effects by tracking fertilizer-derived nitrogen through soil, plant and microbial pools.

Numerous studies employing ^{15}N-urea or $^{15}NH_4^+$ have demonstrated increased recovery of fertilizer nitrogen in plant biomass following biochar application, often accompanied by reduced nitrogen losses via volatilization or leaching [18]. Isotopic mass balance approaches enable the partitioning of applied nitrogen among plant uptake, soil retention, and gaseous losses, revealing that biochar frequently enhances nitrogen use efficiency rather than simply increasing total nitrogen availability [19].

Isotopic techniques are equally critical for evaluating the effects of biochar on biological nitrogen fixation. Using either ^{15}N enrichment or natural abundance methods, researchers have shown that biochar can enhance symbiotic nitrogen fixation in legumes by improving soil physical conditions, nutrient availability and microbial habitat quality [20]. These effects are particularly relevant in low-input or tropical systems, where nitrogen availability often limits productivity.

6. Phosphorus dynamics revealed by radioisotopic techniques

Phosphorus availability is one of the most contentious aspects of biochar research, with reported effects ranging from strong increases to negligible or even negative responses [21]. Radioisotopic tracers, especially ^{32}P and ^{33}P, offer a mechanistic framework to resolve these inconsistencies. By quantifying isotopically exchangeable P pools through dilution techniques, it becomes possible to distinguish between changes in total P content and changes in P availability [22].

In highly weathered tropical soils, biochar-induced increases in soil pH and reductions in active Fe and Al can significantly increase the E-value, reflecting enhanced P availability [23]. Isotopic exchange kinetics further reveal whether biochar introduces fast- or slow-exchanging P pools, information that is inaccessible through routine soil tests. Plant uptake using ^{32}P-labeled fertilizers demonstrate that biochar can increase the proportion of fertilizer-derived P recovered in plant tissue, particularly under P-limiting conditions.

7. Sulfur and micronutrient cycling traced by isotopes

Sulfur dynamics in biochar-amended soils have received comparatively less attention, yet isotopic approaches using ^{35}S and ^{34}S provide valuable insights into sulfate retention, organic S mineralization and microbial immobilization. Biochar-induced changes in redox conditions and microbial activity can substantially alter sulfur cycling [24], effects that are readily captured using isotopic tracers. Similarly, micronutrient dynamics, particularly for Zn, Fe, and Ca, are strongly influenced by biochar's sorptive capacity [25]. The use of γ-emitting isotopes such as ^{65}Zn enables non-destructive tracking of micronutrient uptake and redistribution within plants, revealing whether biochar acts primarily as a sink or as a mediator enhancing nutrient availability under specific soil conditions.

8. Isotopic tracing of contaminant: Biochar interactions

Beyond fertility, biochar is increasingly used to mitigate the mobility and bioavailability of organic contaminants and heavy metals. Radiolabeled compounds such as ^{14}C-pesticides have been instrumental in quantifying sorption strength, desorption kinetics and microbial degradation pathways in biochar-amended soils. These studies demonstrate that biochar can substantially reduce contaminant mobility and leaching while simultaneously altering degradation dynamics, with implications for both agronomic safety and environmental protection.

The broad structural and chemical diversity of organic contaminants and biochar, combined with the many methods used to produce biochar, creates a wide range of possible interactions between these molecules [26]. This leads to different interaction mechanisms that are not yet fully clarified. These interactions may be physical, chemical or a combination of both and together give rise to sorption. The dominant mechanism depends on the interplay between the properties and structural features of both materials (such as porosity, specific surface area, hydrophobicity, ion-exchange capacity, functional groups and aromaticity) and the conditions of the surrounding environment (including concentration, pH and temperature) [26]. Predicting the most likely mechanisms in a sorption system requires integrating knowledge of intermolecular interactions, thermodynamic principles and the structural and compositional characteristics of the substances involved.

The presence of biochars in soils has provided good sorption of many organic contaminants, influencing the mobility of these molecules applied to the soil profile. The effect of biochar application on the transport of ^{14}C-hexazinone, ^{14}C-metribuzin and ^{14}C-quinclorac in soil was demonstrated by Mendes et al. [27]. Using thin-layer chromatography (TLC) soil plates with soil (**Figure 2**), the authors demonstrated that the most significant reductions in herbicide mobility occurred when biochar was incorporated into the soil compared to treatments where biochars were applied in the middle or on top of the TLC soil.

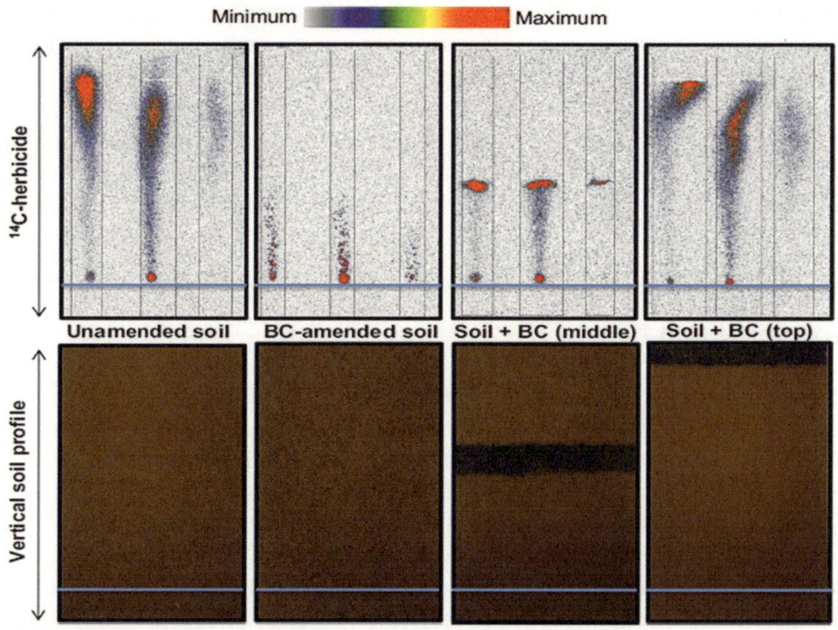

Figure 2.
Mobility of ^{14}C-labeled ^{14}C-quinclorac, ^{14}C-metribuzin and ^{14}C-hexazinone (left to right, respectively) in four soil treatments in TLC-soil: unamended soil (control), soil amended with biochar, soil + biochar strip (1 cm) in the middle, and soil + biochar band at the top of the TLC-soil. Source: Adapted from Mendes et al. [27].

9. Perspectives and future directions

The integration of isotopic techniques into biochar research remains substantially underexplored, particularly when considering the complexity of soil-plant-microbe interactions and the multiple biogeochemical processes simultaneously affected by biochar amendment. While many studies have documented changes in nutrient availability or carbon stocks following biochar application, far fewer have employed isotopic tracers to disentangle the underlying mechanisms governing nutrient fluxes, microbial turnover and element stabilization. As a result, there is a pressing need to move beyond bulk measurements and adopt isotopic frameworks capable of resolving source-sink relationships, turnover rates and transformation pathways within biochar-enriched soil systems.

Emerging methodological advances offer significant opportunities to address these knowledge gaps. Approaches such as compound-specific isotope analysis (CSIA) enable the tracking of individual molecular fractions of organic carbon and nitrogen, providing insights into preferential substrate utilization and microbial processing at the molecular level. Similarly, dual- and multi-isotope labeling strategies, for example, combining ^{13}C or ^{14}C with ^{15}N, ^{33}P or ^{35}S allow simultaneous assessment of coupled biogeochemical cycles and their responses to biochar amendment. When integrated with long-term field experiments and supported by isotopic and process-based modeling, these techniques can bridge the gap between short-term laboratory observations and ecosystem-scale responses, capturing legacy effects of biochar on soil fertility.

Looking forward, the strategic incorporation of isotopic tools into biochar research is essential for developing predictive and transferable knowledge that supports both agronomic and environmental decision-making. Isotopically informed studies can refine fertilizer management strategies, improve assessments of carbon sequestration potential, and enhance the understanding of contaminant fate in biochar-amended soils under realistic field conditions. Ultimately, advancing the integration of biochar science with isotopic methodologies will be critical for translating experimental findings into robust, science-based recommendations and for ensuring that biochar applications contribute effectively to sustainable soil management and environmental protection.

10. Concluding remarks

The body of evidence synthesized in this chapter demonstrates that the integration of biochar research with isotopic methodologies represents a decisive advance in the mechanistic understanding of soil fertility and plant nutrition processes. While numerous studies have reported changes in soil properties and crop performance following biochar application, isotopic tracers uniquely enable the quantification of element fluxes, source contributions, and transformation pathways that underpin these responses. By moving beyond static assessments of nutrient pools, isotopic approaches provide a dynamic and process-oriented framework capable of resolving how biochar alters carbon stabilization, nutrient availability and microbial turnover within soil-plant systems.

Carbon cycling emerges as a central process through which biochar exerts long-term effects, and the combined use of ^{13}C natural abundance and ^{14}C radiotracers has proven particularly powerful in disentangling biochar-derived carbon from native soil organic matter. These approaches reveal that biochar is not an inert amendment, but rather an active component of soil carbon dynamics, capable of inducing priming effects that depend on biochar properties, soil conditions and the availability of labile carbon substrates. Such insights fundamentally refine current paradigms of carbon sequestration by demonstrating that biochar-mediated carbon stabilization must be interpreted within the broader context of soil organic matter turnover and microbial activity.

Beyond carbon, isotopic applications to nitrogen, phosphorus, sulfur and micronutrient cycling highlight the multifaceted role of biochar in regulating nutrient use efficiency rather than simply increasing nutrient stocks. Tracers such as ^{14}C, ^{15}N, ^{32}P, ^{33}P and ^{35}S reveal that biochar can modify nutrient exchange kinetics, microbial immobilization and plant uptake pathways in ways that are highly context-dependent. Importantly, isotopic evidence indicates that improvements in crop nutrition associated with biochar often arise from enhanced synchronization between nutrient supply and plant demand, underscoring the relevance of isotopic tools for designing more efficient and environmentally sound fertilization strategies.

Finally, radiolabeled studies demonstrate that biochar can substantially reduce the mobility and bioavailability of organic contaminants while simultaneously influencing degradation pathways, with direct implications for agrochemical management and soil protection. Looking forward, the systematic incorporation of stable and radioactive isotopes into biochar research, particularly through multi-isotope labeling, compound-specific analyses and long-term field studies, will be essential for generating predictive, transferable knowledge. Such integration is critical for translating experimental findings into robust agronomic recommendations and for ensuring that biochar applications contribute effectively to resilient agricultural systems and environmental protection.

Acknowledgements

This research was supported by the Brazilian National Council for Scientific and Technological Development (CNPq - 151130/2024-8). The author acknowledges the use of QuillBot and SciSpace for language polishing of the manuscript.

Author details

Rodrigo Nogueira de Sousa, Giovani Rangel Malerba Alfonso and Kassio Ferreira Mendes*
Laboratory of Ecotoxicology, Center for Nuclear Energy in Agriculture, University of São Paulo, Piracicaba, SP, Brazil

*Address all correspondence to: kassio_mendes_06@hotmail.com

References

[1] Kuryntseva P, Karamova K, Galitskaya P, Selivanovskaya S, Evtugyn G. Biochar functions in soil depending on feedstock and pyrolyzation properties with particular emphasis on biological properties. Agriculture. 2023;**13**(10):1-39. DOI: 10.3390/agriculture13102003

[2] van Bavel CHM. Water relations of plants and soils. Soil Science. 1996;**161**(4):257-260. DOI: 10.1097/00010694-199604000-00007

[3] Gorovtsov AV, Minkina TM, Mandzhieva SS, et al. The mechanisms of biochar interactions with microorganisms in soil. Environmental Geochemistry and Health. 2020;**42**:2495-2518. DOI: 10.1007/s10653-019-00412-5

[4] Binkley D, Vitousek P. Soil nutrient availability. In: Plant Physiological Ecology: Field Methods and Instrumentation. Dordrecht: Springer Netherlands; 1989. pp. 75-96

[5] Shrivastava M, Srivastava PC, D'Souza SF. Isotopes and tracer techniques for soil analysis. In: Rakshit A, Ghosh S, Chakraborty S, Philip V, Datta A, editors. Soil Analysis: Recent Trends and Applications. Singapore: Springer; 2020. DOI: 10.1007/978-981-15-2039-6_5

[6] Arnikar HJ. Essentials of Nuclear Chemistry (No. 1653). New Delhi: New Age International; 1995

[7] Glibert PM, Middelburg JJ, McClelland JW, Zanden MJV. Stable isotope tracers: Enriching our perspectives and questions on sources, fates, rates, and pathways of major elements in aquatic systems. Limnology and Oceanography. 2018;**64**(3):950-981. DOI: 10.1002/lno.11087

[8] Torn MS, Vitousek PM, Trumbore SE. The influence of nutrient availability on soil organic matter turnover estimated by incubations and radiocarbon modeling. Ecosystems. 2005;**8**:352-372. DOI: 10.1007/s10021-004-0259-8

[9] Nguyen L, Zapata F, Adu-Gyamfi J. The use of phosphorus radioisotopes to investigate soil phosphorus dynamics and cycling in soil-plant systems. In: Soil Phosphorus. Boca Raton: CRC Press; 2016. pp. 285-312

[10] Chalk PM, Craswell ET. An overview of the role and significance of 15N methodologies in quantifying biological N2 fixation (BNF) and BNF dynamics in agro-ecosystems. Symbiosis. 2018;**75**:1-16. DOI: 10.1007/s13199-017-0526-z

[11] Vernooij R, Dusek U, Popa ME, Yao P, Shaikat A, Qiu C, et al. Stable carbon isotopic composition of biomass burning emissions – Implications for estimating the contribution of C3 and C4 plants. Atmospheric Chemistry and Physics. 2022;**22**(4):2871-2890. DOI: 10.5194/acp-22-2871-2022

[12] Chalk P, Smith CJ. 13 C methodologies for quantifying biochar stability in soil: A critique. European Journal of Soil Science. 2022;**73**(3):1-12. DOI: 10.1111/ejss.13245

[13] Peng Y, Zhang B, Guan C, Jiang X, Tan J, Li X. Identifying biotic and abiotic processes of reversing biochar-induced soil phosphorus leaching through biochar modification with MgAl layered (hydr)oxides. The Science of Total Environment. 2022;**843**:157037. DOI: 10.1016/j.scitotenv.2022.157037

[14] Palansooriya KN, Wong JTF, Hashimoto Y, et al. Response of microbial communities to biochar-amended soils: A critical review. Biochar. 2019;**1**:3-22. DOI: 10.1007/s42773-019-00009-2

[15] Kerré B, Hernandez-Soriano MC, Smolders E. Partitioning of carbon sources among functional pools to investigate short-term priming effects of biochar in soil: A 13C study. The Science of the Total Environment. 2016;**547**:30-38. DOI: 10.1016/j.scitotenv.2015.12.107

[16] Luo Y, Zang H, Yu Z, Chen Z, Gunina A, Kuzyakov Y, et al. Priming effects in biochar enriched soils using a three-source-partitioning approach: 14C labelling and 13C natural abundance. Soil Biology and Biochemistry. 2016;**106**: 28-35. DOI: 10.1016/j.soilbio.2016.12.006

[17] Clough TJ, Condron LM. Biochar and the nitrogen cycle: Introduction. Journal of Environmental Quality. 2010;**39**(4):1218-1223. DOI: 10.2134/jeq2010.0204

[18] Alami NE, Bourguid M, Kouighat M. Advances in nitrogen use efficiency of 15N-enriched fertilizers: A thirty-year review of research and innovations for sustainable agriculture. Journal of Soil Science and Plant Nutrition. 2025;**25**:5202-5224. DOI: 10.1007/s42729-025-02458-0

[19] Hood-Nowotny R. The use of stable isotopes in understanding the impact of biochar on the nitrogen cycle. Biochar. 2016;**351**:351-367

[20] Farhangi-Abriz S, Ghassemi-Golezani K, Torabian S, Qin R. A meta-analysis to estimate the potential of biochar in improving nitrogen fixation and plant biomass of legumes. Biomass Conversion and Biorefinery. 2022;**14**(3):3293-3303. DOI: 10.1007/s13399-022-02530-0

[21] Glaser B, Lehr V. Biochar effects on phosphorus availability in agricultural soils: A meta-analysis. Scientific Reports. 2019;**9**(1):9338. DOI: 10.1038/s41598-019-45693-z

[22] Oehl F, Oberson A, Sinaj S, Frossard E. Organic phosphorus mineralization studies using isotopic dilution techniques. Soil Science Society of America Journal. 2001;**65**(3):780-787. DOI: 10.2136/sssaj2001.653780x

[23] Yu J. Characterization of Tropical Agricultural Soil Microbiomes after Biochar Amendment [Doctoral dissertation]. Tempe: Arizona State University; 2020

[24] Wang Y, Zhang Y, Ok YS, Jiang T, Liu P, Shu R, et al. Biochar-impacted sulfur cycling affects methylmercury phytoavailability in soils under different redox conditions. Journal of Hazardous Materials. 2020;**407**:124397. DOI: 10.1016/j.jhazmat.2020.124397

[25] Hossain MZ, Bahar MM, Sarkar B, et al. Biochar and its importance on nutrient dynamics in soil and plant. Biochar. 2020;**2**:379-420. DOI: 10.1007/s42773-020-00065-z

[26] Sousa RN, Soares MB, Santos FH, Leite CN, Mendes KF. Interaction mechanisms between biochar and herbicides. In: Mendes KF, editor. Interactions of Biochar and Herbicides in the Environment. 1st ed. Vol. 1. Boca Raton: CRC Press; 2022. pp. 133-214. DOI: 10.1201/9781003202073-4

[27] Mendes KF, Sousa RN, Takeshita V, Alonso FG, Régo APJ, Tornisielo VL. Cow bone char as a sorbent to increase sorption and decrease mobility of hexazinone, metribuzin, and quinclorac in soil. Geoderma. 2019;**343**:40-49. DOI: 10.1016/j.geoderma

Section 2

Biochar Applications in Agriculture, Soil Health and Plant Protection

Chapter 5

Research and Application Progress of Biochar in Amelioration of Saline-Alkali Soil

Peng Zhang, Yuning Wang, Fude Liu, Le Jiao and Hongwen Sun

Abstract

Saline-alkali soils pose a significant global challenge to agricultural productivity, necessitating sustainable remediation strategies to unlock their cultivation potential. Biochar has emerged as a promising soil amendment, exhibiting multifunctional capabilities in pH regulation, contaminant immobilization, carbon sequestration, and nutrient retention. However, the heterogeneous nature of biochar compositions results in inconsistent remediation outcomes, while the mechanistic understanding of its interactions with saline-alkali soils remains incomplete. This chapter provides a comprehensive analysis of the effects of biochar on the physicochemical properties, nutrient availability, and biological characteristics of saline-alkali soils. It summarizes the improvement effects of both biochar and modified biochar on saline-alkali soils, evaluates their impacts on quality and efficiency, and elucidates the potential mechanisms underlying biochar's role in saline-alkali soil remediation. Furthermore, this chapter outlines future research directions to provide a reference for developing green, efficient, and precise biochar-based technologies tailored to saline-alkali soil remediation, bridging the gap between fundamental research and large-scale agricultural applications.

Keywords: saline-alkali soil, biochar, modification, physicochemical property, nutrient

1. Introduction

Soil salinization is recognized as one of the most significant challenges threatening agricultural production, food security, and sustainable development in arid and semi-arid regions globally. Currently, the global saline-alkali land area is approximately 1.1×10^9 hm^2 (accounting for 8.7% of the Earth's total land area) [1]. These areas are predominantly located in naturally arid or semi-arid regions of Asia, Africa, and Latin America. Saline-alkali lands are typically characterized by a scarcity of freshwater resources, low terrain, high groundwater levels, and difficulties in soil desalination [2]. The issues of declining soil fertility, poor crop growth, and fragile ecological environments caused by soil salinization severely constrain regional agricultural, forestry, and urban development. Developing soil conditioners that are highly compatible with traditional physical, chemical, and biological improvement techniques, as well as engineering measures, represents a key strategy for the green, efficient, and precise improvement and comprehensive utilization of saline-alkali land.

In recent years, biochar technology has emerged as a research hotspot in the fields of soil science and environmental science. It is considered an important "multi-functional" technology capable of addressing major challenges affecting human sustainable development, such as climate change [3], agricultural production safety [4], energy crises [5], agricultural waste resource utilization [6], and environmental pollution [7]. This has sparked a global research and development boom. Biochar is a highly aromatic carbonaceous material produced through the pyrolysis and carbonization of animal and plant residues under completely or partially oxygen-deficient conditions. It exhibits characteristics such as a large specific surface area, strong adsorption capacity, and anti-decomposition properties [5], and plays positive roles in soil improvement, crop yield enhancement, and environmental protection. Extensive studies have demonstrated that biochar exerts direct or indirect regulatory effects on soil nutrients [8], increases soil organic matter content, cation exchange capacity (CEC), permeability, and microbial activity [9], reduces nitrogen and phosphorus losses, enhances soil aggregate structure formation [10], and improves soil water conductivity to facilitate water infiltration, thereby optimizing crop habitats and increasing yields. Over the past two decades, biochar technology has achieved notable progress in the improvement and application of saline-alkali land. Numerous studies have confirmed the beneficial effects of biochar application on saline-alkali soil properties, including enhancing soil permeability and organic matter content [9], reducing pH and salinity [11], and promoting microbial and crop growth [12]. However, due to the high salt ion content and strong alkalinity of biochar produced at large scales, some studies have also reported negative impacts, such as increased soil pH and salt content [13]. These findings indicate that current biochar technology exhibits contradictions and lacks systematicness in saline-alkali land improvement, which hinders accurate prediction and evaluation of its scientific and practical value for different types of saline-alkali land in China. Additionally, due to the complex interaction processes between biochar and soil, some research results remain superficial, and the elucidation of microscopic mechanisms requires further refinement. Therefore, as a novel comprehensive remediation technology for saline-alkali land, a comprehensive and systematic evaluation of the application value of biochar technology in saline-alkali land improvement, clarification of relevant processes and micro-mechanisms, and identification of its advantages and disadvantages can provide correct guidance for the widespread application of biochar technology in saline-alkali land improvement in China and promote the healthy development of biochar-based improvement technologies.

Based on this, this chapter systematically reviews the research status of biochar in saline-alkali land improvement over the past 20 years, analyzes and summarizes the influence of biochar on the physical and chemical properties, nutrient availability, and biological characteristics of saline-alkali soils, and elucidates the key mechanisms of biochar improvement and its potential application value in combination with the unique characteristics of biochar. Finally, in light of existing problems and deficiencies, future research directions and prospects are proposed. This study aims to provide a theoretical basis for promoting and applying biochar technology in saline-alkali land improvement and contribute to the exploration and development of green, low-carbon, environmentally friendly, and sustainable saline-alkali land improvement technologies.

2. Effect of biochar on the physicochemical properties of saline-alkali soil

Biochar exhibits a high specific surface area, complex pore structure, and abundant functional groups. These characteristics enable biochar to play multiple roles in the improvement of saline-alkali land [14]. The physicochemical properties of biochar are influenced by various factors, including the type of raw materials, pyrolysis temperature, preparation method, heating rate, and residence time [15]. These factors also significantly affect the promoting effects of biochar on the physicochemical properties of saline-alkali soil, water and salt transport and distribution, as well as agricultural productivity. Based on numerous studies, the addition of biochar prepared from different raw materials and pyrolyzed under varying conditions has diverse impacts on soil properties, such as soluble salt content, pH, bulk density, CEC, organic matter content, and crop growth and yield in different types of saline-alkali soils.

Total soil salinity and degree of alkalization (exchangeable sodium percentage (ESP)) are critical indicators for assessing the extent of soil salinization and alkalization. Saline-alkali soils are typically characterized by high salt content, high electrical conductivity (EC), and elevated ESP due to the accumulation of large amounts of salts. Reducing the soluble salt content and ESP in the soil, alleviating salt stress on soil organisms, and restoring soil ecological functions remain the core objectives of saline-alkali land improvement. In the early stages of biochar technology development, biochar was primarily produced through high-temperature pyrolysis, resulting in materials rich in alkali metal ions and exhibiting strong alkalinity. Consequently, earlier studies predominantly focused on its application in improving acidic soils. However, in recent years, an increasing number of studies [16, 17] have highlighted the potential of biochar in mitigating the adverse effects of saline-alkali soils. These studies confirm that biochar can effectively reduce the salt content and ESP of saline-alkali soils. The mechanisms underlying biochar's ability to reduce soil salinity and ESP in saline-alkali soils can be attributed to two primary aspects: (1) The biochar surface is enriched with exchangeable ions such as Ca^{2+} and Mg^{2+}, which can displace adsorbed Na^{+} from soil colloids, thereby reducing sodium salt content in the soil [18]; and (2) Biochar significantly enhances soil porosity and specific surface area, improves soil water retention capacity, and facilitates the leaching of sodium salts from the soil [19, 20].

Biochar can alter the soil pH of saline-alkali land, primarily attributed to the leaching of substances such as alkali metal ions and carbonates that are abundant in biochar. However, these changes are influenced by the type of biochar raw materials and pyrolysis temperature. For example, Sun et al. [21] reported that when wheat straw biochar (pH = 6.9), corn straw biochar (pH = 8.0), and peanut shell biochar (pH = 7.7) were added to coastal sodium chloride saline-alkali soil (pH = 8.6), only wheat straw biochar significantly increased the soil pH, while no significant changes were observed with the other two types. Further studies have confirmed that as pyrolysis temperature increases, the ash content of biochar rises, leading to enhanced alkalinity [22]. Conversely, biochar produced at low pyrolysis temperatures (< 500°C) tends to be neutral or weakly alkaline, making it more suitable for improving saline-alkali soils [18]. Additionally, research has shown that biochar produced via hydrothermal methods exhibits acidity and can effectively reduce soil pH [23]. However, this preparation method imposes high demands on equipment, involves significant technical challenges, and exhibits poor safety performance, thereby limiting its potential for large-scale production. In summary, the effectiveness of biochar in altering the soil pH of saline-alkali land is constrained by multiple factors, including soil type, biochar raw materials, pyrolysis temperature, biochar properties, and application methods.

The porosity of saline-alkali soil is generally low [1], which restricts plant growth. Numerous studies [24, 25] have demonstrated that biochar can significantly enhance the porosity of saline-alkali soil, particularly in the topsoil layer (0–30 cm), with increases ranging from 4.8 to 30.1%. This effect is directly attributed to the high intrinsic porosity of biochar, which, upon incorporation into the soil, directly elevates the overall soil porosity [26]. Additionally, the poor physical conditions of saline-alkali soil, such as low porosity and high bulk density, are mitigated by biochar's large specific surface area, low density, stable porous structure, and significantly lower bulk density compared to saline-alkali soil [27]. Biochar not only improves soil aggregate structure and enhances aggregate stability but also increases soil porosity while containing substantial organic matter that dilutes soil mineral components [24], indirectly reducing soil bulk density. Therefore, for improving the porosity and bulk density of saline-alkali soil, biochar with low bulk density, high porosity, and low ash content should be prioritized, such as straw-derived biochar produced at high pyrolysis temperatures (> 600°C) [28, 29].

Saline-alkali soils exhibit low organic matter content, and their exchange sites are predominantly occupied by salt ions such as Na^+, limiting the soil's capacity to retain other cations and resulting in a low CEC. Consequently, CEC serves as a critical indicator for evaluating the improvement of saline-alkali soils. Extensive research has confirmed that biochar, due to its abundance of functional groups (e.g., -OH and -COOH), enhances its adsorption capacity for soil cations, effectively increasing the CEC of saline-alkali soils by 22.5–82.0% [30, 31]. The mechanisms underlying this enhancement include: (1) Biochar's high specific surface area and rich oxygen-containing functional groups (e.g., carboxyl and hydroxyl groups) and metal minerals directly increase soil exchange sites [18]; and (2) The pH increase induced by biochar leads to soil colloid deprotonation, promoting an increase in soil CEC [32]. The determination of biochar's CEC value hinges on its surface oxygen-containing functional groups and metal mineral composition, leading to variations in CEC values among biochars prepared from different raw materials at varying temperatures [33]. Currently, limited research exists on the mechanisms of biochar's effects on saline-alkali soil CEC, particularly regarding the interplay between environmental factors (e.g., flooding and drought), soil texture, biochar properties, and saline-alkali soil improvement, necessitating further long-term field experiments for clarification. Soil carbon constitutes the largest terrestrial carbon pool, and even minor changes in soil carbon can significantly impact atmospheric CO_2 concentrations and the global carbon cycle. As a soil amendment rich in organic carbon, biochar has been shown in numerous studies to enhance carbon sequestration in saline-alkali soils [34] and reduce greenhouse gas emissions [35]. Upon entering the soil, biochar exhibits two primary mechanisms: carbon sequestration (increasing carbon sinks) and enhanced carbon emissions. These mechanisms influence the exchange of three major greenhouse gases (CO_2, CH_4, and N_2O) [36] between farmland systems and the atmosphere. Whether biochar acts as a net carbon sink depends on the dominance of these mechanisms. Key carbon sequestration mechanisms include: (1) Biochar's high proportion of recalcitrant carbon enables long-term sequestration within the soil, forming a stable carbon reservoir [37]; (2) Its large specific surface area facilitates the adsorption of organic and inorganic substances containing C and N, generating more stable organic–inorganic complexes and reducing carbon decomposition and release [38]; and (3) By regulating soil properties, inhibiting microbial activity, and reducing enzyme activity, biochar influences soil carbon and nitrogen cycles, thereby modulating soil respiration, nitrification, and denitrification reactions to mitigate greenhouse gas emissions [39]. Overall, biochar represents an effective amendment for enhancing soil organic carbon and carbon sink levels in saline-alkali land. Future research should focus on elucidating the regulatory mechanisms of biochar on soil carbon sequestration and carbon emissions in saline-alkali land, further exploring biochar's potential for soil carbon storage, and providing theoretical and technical support for precise regulation of soil carbon cycles in saline-alkali land.

3. Effects of biochar on nutrient availability in saline-alkali soil

Due to the poor soil structure in saline-alkali soils, anions such as Cl^-, SO_4^{2-}, and OH^- compete with nutrients like N and P for adsorption sites on soil particles [40], leading to the easy loss of soil nutrients through water flow or leaching. Additionally, the scarcity of available nutrients in saline-alkali soils results in extremely low nutrient utilization rates by plants [41]. Biochar is inherently rich in essential nutrients such as N, P, K, and minerals, and can significantly enhance the nutrient content in saline-alkali soils. The primary mechanisms are as follows: (1) Directly increasing soil nutrient levels through its inherent nutrient content [42]. For instance, biochar derived from livestock and poultry manure exhibits relatively high nitrogen content [43]. Moreover, the water-soluble nitrogen and phosphorus content in biochar gradually decreases with increasing pyrolysis temperature [44–46]. And as pyrolysis temperature increases, inorganic orthophosphate (PO_4^{3-}) becomes the sole form of phosphorus in manure-derived biochar at 500°C, while water-soluble phosphorus is nearly absent at high temperatures (≥ 700°C). Therefore, low-temperature biochar derived from livestock and poultry manure is a more suitable choice for enhancing nutrient content in saline-alkali soils. (2) Biochar enhances soil aggregate stability [47], thereby improving the soil's capacity to retain nutrients and reducing nutrient leaching [8]. (3) Biochar alleviates salt stress in crops and promotes the absorption of free nutrients in the soil, which benefits crop growth [48].

Simultaneously, biochar influences the cycling of nitrogen (N), phosphorus (P), and potassium (K) (NPK) elements in the soil by enhancing microbial activity and soil enzyme activity, thereby reducing nutrient losses [41]. For example, biochar reduces ammonia (NH_3) and nitrous oxide (N_2O) emissions [49], minimizes NO_3^- leaching [47], promotes nitrification, and enhances nitrogen fixation [50], thus increasing the proportion of plant-available N in the soil. Biochar also increases phosphatase activity [9] and the abundance and activity of rhizosphere microorganisms [51], promoting the mineralization of organic phosphorus and converting it into plant-available forms, further enhancing phosphorus availability. The behavior of K in soil differs from those of N and P, as K dynamics are not influenced by microbial activity but are primarily governed by the cation exchange characteristics of the soil [52]. Currently, there is limited research on the mechanisms by which biochar affects potassium in saline-alkali soils; however, some studies indicate that biochar application significantly increases soil CEC and available potassium content [53].

4. Effects of biochar on soil biological properties in saline-alkali soil

The capacity of saline-alkali land to support plants, animals, microbial abundance, and enzyme activity is significantly lower than that of normal soil [54]. Numerous studies have investigated the effects of biochar addition on the structure and abundance of soil microbial communities in saline-alkali land. Research has demonstrated that due to its high specific surface area, porous structure, and rich content of organic matter, N, P, K, and other nutrients, biochar can increase the levels of soluble organic matter and inorganic nutrients in soil, reduce salt content, alleviate salt stress, and thereby enhance microbial abundance and enzyme activity [55, 56]. However, variations in raw materials and preparation processes result in significant differences in the pH and nutrient content of biochar. Tang et al. [22] found that high-temperature biochar (> 550°C) contains fewer toxic functional groups, such as carboxylic acids, phenols, and amines, which promote the transformation of soil microbial communities toward beneficial bacterial groups, such as Proteobacteria and Bacteroidetes. Additionally, the type and application rate of biochar can alter the activity of soil enzymes, such as sucrase, urease, and phosphatase. The application rate of biochar also affects soil enzyme activity; when the application rate is less than 2%, soil enzyme activity increases rapidly, whereas at rates greater than 2%, the increase becomes slower [57].

The community structure and abundance of soil fauna are closely linked to the physicochemical properties of saline-alkali soil. For example, Zhang Jianying et al. [58] reported that the distribution of groups, such as Formicidae, Muridae, and Curculionidae, is negatively correlated with soil pH, alkalinity, and salt content. Soil factors such as total salt and total nitrogen play a decisive role in shaping the composition of soil animal communities in saline-alkali land. By reducing soil salt content and improving soil physicochemical properties, biochar provides a more suitable habitat for soil fauna. In contrast to the extensive research on soil microorganisms and enzymes, there are relatively few studies on the interactions between biochar and small-to-medium-sized soil fauna such as earthworms in saline-alkali land. Some studies have focused on the inherent ecological risks of biochar, such as polycyclic aromatic hydrocarbons (PAHs) and persistent free radicals, which may be toxic to soil fauna and inhibit their growth and reproduction [29]. These negative effects of biochar on soil fauna indirectly highlight the ecological risks associated with biochar use in saline-alkali land remediation. Therefore, ensuring the ecological safety of biochar is an urgent issue that must be addressed in future saline-alkali land management strategies.

For terrestrial plants, growing in saline-alkali land presents significant challenges. First, high salt stress in soil solution directly inhibits plant growth. Second, poor soil structure and severe deficiencies in air and water lead to underdeveloped root systems and low emergence rates. Finally, nutrient scarcity results in poor plant nutrition. As discussed earlier, biochar performs effectively in addressing high salt content, poor soil structure, and low nutrient availability, providing a favorable growth environment for plants. However, persistent free radicals in biochar [59] can accelerate plant cell aging, disrupt the growth and metabolic cycles of epidermal cells, and inhibit plant development. Furthermore, high application rates of biochar [60] and increased soil pH and salt content following biochar application may negatively affect plant growth and development. Therefore, future research should focus on elucidating the intrinsic effects and mechanisms of biochar on crops and prioritize the development of novel biochar types with low salt, low toxicity, and high efficiency for saline-alkali land improvement.

5. Amelioration effect of modified biochar for saline-alkaline soil

Although biochar offers numerous advantages for the amelioration of saline-alkaline soil, its single application not only requires large quantities and incurs high costs but may also increase the soil pH and salt content, negatively impacting saline-alkaline soil. To address these issues, various biochar modification techniques have been proposed to enhance quality and efficiency, including acid washing and activation, loading active components, and physical mixing with traditional fertilizers or soil amendments [61, 62]. Modified biochar exhibits advantages that primary biochar cannot match, significantly enhancing its amelioration effects on saline-alkaline soil [63, 64].

To overcome the strong alkalinity of conventional pyrolyzed biochar, researchers have developed methods involving treatment with strong acids such as nitric acid and hydrochloric acid, known as acid-modified biochar [65, 66]. After acid treatment, inorganic components and organic matter leach from the surface and pores of biochar, while alkaline components are removed or neutralized. This process exposes more micropores, increases the specific surface area [67], and enhances oxygen-containing functional groups [68], thereby strengthening biochar's adsorption capacity for soil salt ions. Additionally, acid-modified biochar significantly improves soil porosity and permeability, increasing the proportion of water-stable large aggregates [28]. Some studies have also demonstrated that preparing biochar via hydrothermal treatment in a strong acid environment [65] can retain more acidic oxygen-containing functional groups and pore structures [69], enhancing its adsorption capacity for soil salt ions [70] and improving its amelioration effects on saline-alkaline soil. Despite extensive research on strong acid modification, challenges remain regarding stability, economic cost, and amelioration efficiency in practical saline-alkaline soil applications, necessitating further evaluation in future studies.

Due to the limited nutrient content in biochar, its single application cannot provide sufficient nutrients for crops [71]. Consequently, researchers have developed biochar-based fertilizers, which can directly supplement required nutrients [72], delay nutrient release in the soil, reduce nutrient loss, and serve as carriers for enhancing fertilizer efficiency [73]. However, the effects and mechanisms of different fertilizer types vary: (1) Organic fertilizers such as poultry manure are rich in organic carbon, increasing the abundance of soil microbial communities in saline-alkaline soil and promoting crop growth [74]. Simultaneously, the clogging of micropores by adsorbed compost-derived substances reduces biochar's surface area, alleviating the negative effects of excessive biochar on soil [75]. (2) Combining biochar with inorganic fertilizers such as nitrogen and phosphorus fertilizers generates higher levels of available nutrients, enhancing soil enzyme activity and microbial abundance in saline-alkaline soil [76]. Moreover, biochar adsorbs nutrients, reducing the leaching of inorganic nitrogen and phosphorus in the soil. Notably, due to the adsorption or precipitation reactions of orthophosphate (PO_4^{3-}) at higher pH values, phosphorus fertilizers and biochar in saline-alkaline soil may exhibit negative interactions [77]. (3) When co-applied with microbial fertilizers, biochar not only provides nutrients to plants as a fertilizer but also creates suitable habitats for microorganisms [78], mitigating the stress caused by harsh saline-alkaline soil conditions [79]. Overall, the application of biochar-based fertilizers significantly enhances biochar's amelioration efficiency on saline-alkaline soil while reducing economic costs. However, further research is needed to evaluate the applicability of biochar-based fertilizers for different types and textures of saline-alkaline soils, providing critical guidance for precise fertilization control and improving biochar utilization efficiency.

The development of new composite materials (amendments) through physical and chemical combinations of biochar and loadings represents a current research hotspot. Due to its rough and porous surface morphology, biochar serves as an excellent carrier medium. Loading substances generate new functional groups and salt ion adsorption sites on biochar's surface. Common loading substances include metal compounds and clay minerals. The mechanisms by which biochar-based composite materials improve saline-alkaline land primarily include those given as follows: (1) Loading substances introduce new elements and functional groups [80], enhancing biochar's complexation and precipitation capacities for soil soluble salt ions, reducing soil soluble salt content and alkalinity. Zhang et al. [81] noted that after biochar was loaded with red phosphorus, new phosphorus-containing functional groups formed on its surface, creating new phosphorus-containing minerals with soil metal ions. This significantly reduced the electrical conductivity and soluble salt content of saline-alkaline soil while increasing plant-available phosphorus content. (2) New substance loading provides additional adsorption sites for soil nutrients such as nitrogen and phosphorus, reducing nutrient leaching [82]. (3) Metal species such as iron, calcium, and magnesium loaded in composite materials release large numbers of metal cations, exchanging adsorbed Na^+ in soil colloids, accelerating Na^+ migration during soil percolation, improving soil leaching effects, and reducing sodium salt content [83]. In summary, biochar-based composite materials integrate the advantages of biochar and loading substances, enabling efficient and precise improvement and governance of saline-alkaline land by tailoring loads to regional saline-alkaline soil types.

6. Conclusions and prospects

Biochar technology has emerged as a research hotspot in soil science and environmental science in recent years, achieving notable progress in the improvement of saline-alkali land. Specifically, biochar exhibits a large specific surface area and strong adsorption capacity, enabling it to directly or indirectly regulate soil nutrients. It enhances soil organic matter content, CEC, permeability, and microbial activity in saline-alkali land, reduces nitrogen and phosphorus losses, and promotes the formation of soil aggregate structures. Biochar plays a positive role in soil improvement, water retention, and crop resistance to salt and alkali stresses in saline-alkali land. Additionally, when biochar is returned to agricultural fields, it expands the soil carbon pool and demonstrates superior carbon sequestration and emission reduction effects compared to other agricultural practices. Biochar-based fertilizers or composite materials used for saline-alkali land improvement can delay fertilizer release, enhance soil fertility persistence, and improve crop adaptability to saline-alkali conditions. However, numerous challenges remain in the research and application of biochar for saline-alkali land improvement, necessitating further investigation:

1. Basic theoretical research on biochar improvement of saline-alkali land remains controversial and uncertain. The impacts of biochar on the physicochemical properties and functions of saline-alkali soils, the geochemical cycles of key elements such as nitrogen and phosphorus, the succession and functions of soil microbial communities, and the underlying regulatory mechanisms are not fully understood. In particular, the correlation between biochar properties and the improvement effects on different types of saline-alkali land requires further systematic research and validation.

2. Most studies on biochar improvement of saline-alkali land are still at the laboratory simulation stage, with a lack of long-term, fixed-point, and large-scale field

experiments to analyze its practical applicability and mechanisms of action. For example, whether the patterns of soil water and salt movement under biochar influence are affected by saline-alkali land type, agronomic measures, and irrigation conditions remains unclear. Additionally, the effects of microorganisms and crop roots on the structure and stability of biochar-soil aggregates require further exploration. Simultaneously, factors, such as biochar aging, climate, geology, and hydrology, should be comprehensively considered. Long-term dynamic tracking and monitoring of biochar, combined with soil quality evaluations, are essential to assess the sustainability of biochar's improvement effects and associated environmental risks.

3. Batch production and large-scale application of modified biochar face significant obstacles. Research on raw material selection and preparation processes is predominantly at the laboratory stage, making it challenging to achieve efficient and cost-effective large-scale production. Moreover, the actual soil environment of saline-alkali land is a complex and diversified system, and the performance of modified biochar in real-world soils is influenced by various conditions. Future research should focus on advancing biochar modification methods and principles, promoting large-scale applications, conducting economic cost assessments, and enhancing environmental safety.

Author details

Peng Zhang[1*], Yuning Wang[1], Fude Liu[2], Le Jiao[3] and Hongwen Sun[1]

1 College of Environmental Science and Engineering, Nankai University, Tianjin, China

2 School of Environment Science and Safety Engineering, Tianjin University of Technology, Tianjin, China

3 Agro-Environmental Protection Institute, Ministry of Agriculture and Rural Affairs, Tianjin, China

*Address all correspondence to: nkzhangpeng@nankai.edu.cn

References

[1] Lin ZQ, Bañuelos GS. Soil salination indicators. Dordrecht: Springer Netherlands; 2015

[2] Zhu JK. Plant salt tolerance. Trends in Plant Science. 2001;**6**(2):66-71. DOI: 10.1016/S1360-1385(00)01838-0

[3] Xiao X, Chen B, Chen Z, et al. Insight into multiple and multilevel structures of biochars and their potential environmental applications: A critical review. Environmental Science and Technology. 2018;**52**(9):5027-5047. DOI: 10.1021/acs.est.7b06487

[4] Das SK, Ghosh GK, Avasthe R. Application of biochar in agriculture and environment, and its safety issues. Biomass Conversion and Biorefinery. 2020;**13**:1359-1369. DOI: 10.1007/s13399-020-01106-0

[5] Awasthi MK. Engineered biochar: A multifunctional material for energy and environment. Environmental Pollution. 2022;**298**:118831. DOI: 10.1016/j.envpol.2021.118831

[6] Yuan P, Wang J, Pan Y, et al. Review of biochar for the management of contaminated soil: Preparation, application and prospect. Science of the Total Environment. 2019;**659**:473-490. DOI: 10.1016/j.scitotenv.2018.12.400

[7] Bolan N, Hoang SA, Beiyuan J, et al. Multifunctional applications of biochar beyond carbon storage. International Materials Reviews. 2022;**67**(2):150-200. DOI: 10.1080/09506608.2021.1922047

[8] Li Y, Li G. Mechanisms of straw biochar's improvement of phosphorus bioavailability in soda saline-alkali soil. Environmental Science and Pollution Research. 2022;**29**(32):47867-47872. DOI: 10.1007/s11356-022-20489

[9] Saifullah DS, Naeem A, et al. Biochar application for the remediation of salt-affected soils: Challenges and opportunities. Science of the Total Environment. 2018;**625**:320-335. DOI: 10.1016/j.scitotenv.2017.12.257

[10] Zhang Y, Wang J, Feng Y. The effects of biochar addition on soil physicochemical properties: A review. Catena. 2021;**202**:105284. DOI: 10.1016/j.catena.2021.105284

[11] Oliveira FR, Patel AK, Jaisi DP, et al. Environmental application of biochar: Current status and perspectives. Bioresource Technology. 2017;**246**:110-122. DOI: 10.1016/j.biortech.2017.07.033

[12] Gopinath KP, Vo D-VN, Gnana Prakash D, et al. Environmental applications of carbon-based materials: A review. Environmental Chemistry Letters. 2020;**19**(1):557-582. DOI: 10.1007/s10311-019-00940-7

[13] Zhang C, Zeng G, Huang D, et al. Biochar for environmental management: Mitigating greenhouse gas emissions, contaminant treatment, and potential negative impacts. Chemical Engineering Journal. 2019;**373**:902-922. DOI: 10.1016/j.cej.2019.04.122

[14] Liu M, Wang ZC, Yang F, et al. Application progress of biochar in amelioration of saline-alkaline soil. Journal of Soil and Water Conservation. 2021;**35**(3):1-8

[15] Uday V, Harikrishnan PS, Deoli K, et al. Current trends in production, morphology, and real-world environmental applications

of biochar for the promotion of sustainability. Bioresource Technology. 2022;**359**:127467. DOI: 10.1016/j.biortech.2022.127467

[16] Zhao B, Nan X, Xu H, et al. Sulfate sorption on rape (*Brassica campestris* L.) straw biochar, loess soil and a biochar-soil mixture. Journal of Environmental Management. 2017;**201**:309-314. DOI: 10.1016/j,jenvman.2017.06.030

[17] Tariqul I, Yanliang L, Hefa C. Biochars and engineered biochars for water and soil remediation: A review. Sustainability. 2021;**13**(17):9932. DOI: 10.3390/su13179932

[18] Abbas G, Abrar MM, Naeem MA, et al. Biochar increases salt tolerance and grain yield of quinoa on saline-sodic soil: Multivariate comparison of physiological and oxidative stress attributes. Journal of Soils and Sediments. 2022;**22**(5):1446-1459. DOI: 10.1007/s11368-022-03206-0

[19] Egamberdieva D, Ma H, Alaylar B, et al. Biochar amendments improve licorice (*Glycyrrhiza uralensis* Fisch.) growth and nutrient uptake under salt stress. Plants (Basel). 2021;**10**(10):2135. DOI: 10.3390/plants10102135

[20] Yue Y, Guo WN, Lin QM, et al. Improving salt leaching in a simulated saline soil column by three biochars derived from rice straw, sunflower straw, and cow manure. Journal of Soil and Water Conservation. 2016;**71**(6):467. DOI: 10.2489/jswc.71.6.467

[21] Sun J, He F, Shao H, et al. Effects of biochar application on Suaeda salsa growth and saline soil properties. Environmental Earth Sciences. 2016;**75**(8):630. DOI: 10.1007/s12665-016-5550-0

[22] Tang J, Zhang S, Zhang X, et al. Effects of pyrolysis temperature on soil-plant-microbe responses to Solidago canadensis L.-derived biochar in coastal saline-alkali soil. Science of the Total Environment. 2020;**731**:138938. DOI: 10.1016/j.scitotenv.2020.138938

[23] Harjeet N, Biswajit S, Sudip M, et al. Biochar from biomass: A review on biochar preparation its modification and impact on soil including soil microbiology. Geomicrobiology Journal. 2022;**39**(3-5):373-388. DOI: 10.1080/01490451.2021.1998043

[24] Liang J, Li Y, Si B, et al. Optimizing biochar application to improve soil physical and hydraulic properties in saline-alkali soils. Science of the Total Environment. 2021;**771**:144802. DOI: 10.1016/j.scitotenv.2021.144802

[25] Artiola J, Rasmussen C, Freitas R. Effects of a biochar-amended alkaline soil on the growth of romaine lettuce and bermudagrass. Soil Science. 2012;**177**:561-570. DOI: 10.1097/SS.0b013e31826d3b3d

[26] Burrell LD, Zehetner F, Rampazzo N, et al. Long-term effects of biochar on soil physical properties. Geoderma. 2016;**282**:96-102. DOI: 10.1016/j.geoderma.2016.06.013

[27] Obia A, Mulder J, Hale SE, et al. The potential of biochar in improving drainage, aeration and maize yields in heavy clay soils. PLoS One. 2018;**13**(5):e0196794. DOI: 10.1371/journal.pone.0196794

[28] Suliman W, Harsh JB, Abu-Lail NI, et al. The role of biochar porosity and surface functionality in augmenting hydrologic properties of a sandy soil. Science of the Total Environment. 2017;**574**:139-147. DOI: 10.1016/j.scitotenv.2016.09.053

[29] Uchimiya M, Wartelle LH, Klasson KT, et al. Influence of pyrolysis temperature on biochar property and function as a heavy metal sorbent in soil. Journal of Agricultural and Food Chemistry. 2011;**59**(6):2501-2510. DOI: 10.1021/jf103934k

[30] Cui L, Liu Y, Yan J, et al. Revitalizing coastal saline-alkali soil with biochar application for improved crop growth. Ecological Engineering. 2022;**179**:106594. DOI: 10.1016/j.ecoleng.2022.106594

[31] Zhou Z, Li Z, Zhang Z, et al. Treatment of the saline-alkali soil with acidic corn stalk biochar and its effect on the sorghum yield in western Songnen plain. Science of the Total Environment. 2021;**797**:149190. DOI: 10.1016/j.scitotenv.2021.149190

[32] Atkinson CJ, Fitzgerald JD, Hipps NA. Potential mechanisms for achieving agricultural benefits from biochar application to temperate soils: A review. Plant and Soil. 2010;**337**(1):1-18. DOI: 10.1007/s11104-010-0558-3

[33] Ahmad M, Rajapaksha AU, Lim JE, et al. Biochar as a sorbent for contaminant management in soil and water: A review. Chemosphere. 2014;**99**:19-33. DOI: 10.1016/j.chemosphere.2013.11.053

[34] Fungo B, Lehmann J, Kalbitz K, et al. Emissions intensity and carbon stocks of a tropical Ultisol after amendment with Tithonia green manure, urea and biochar. Field Crops Research. 2017;**209**:179-188. DOI: 10.1016/j.fcr.2017.05.013

[35] Thers H, Djomo SN, Elsgaard L, et al. Biochar potentially mitigates greenhouse gas emissions from cultivation of oilseed rape for biodiesel. Science of the Total Environment. 2019;**671**:180-188. DOI: 10.1016/j.scitotenv.2019.03.423

[36] Wang L, Deng J, Yang X, et al. Role of biochar toward carbon neutrality. Carbon Research. 2023;**2**(1):2. DOI: 10.1007/s44244-023-00035-5

[37] Mukherjee A, Lal R. The biochar dilemma. Soil Research. 2014;**52**(3):217-230. DOI: 10.1071/SR13173

[38] Chen M, Zhang S, Liu L, et al. Organic fertilization increased soil organic carbon stability and sequestration by improving aggregate stability and iron oxide transformation in saline-alkaline soil. Plant and Soil. 2022;**474**(1-2):233-249. DOI: 10.1007/s11104-021-05223-0

[39] Li Y, Hu S, Chen J, et al. Effects of biochar application in forest ecosystems on soil properties and greenhouse gas emissions: A review. Journal of Soils and Sediments. 2018;**18**(2):546-563. DOI: 10.1007/s11368-017-1783-2. DOI: 10.1007/s11368-017-1783-2

[40] Zhang J, Zhou S, Sun H, et al. Three-year rice grain yield responses to coastal mudflat soil properties amended with straw biochar. Journal of Environmental Management. 2019;**239**:23-29. DOI: 10.1016/j.jenvman.2019.03.033

[41] Wu LP, Zhang SR, Ma M, et al. Carbon sequestration under different organic amendments in saline-alkaline soils. Catena. 2020;**196**:104882. DOI: 10.1016/j.catena.2020.104882

[42] Jindo K, Audette Y, Higashikawa FS, et al. Role of biochar in promoting circular economy in the agriculture sector. Part 1: A review of the biochar roles in soil N, P and K cycles. Chemical and Biological Technologies in Agriculture. 2020;**7**(1):15. DOI: 10.1186/s40538-020-0133-0

[43] Al-Wabel M, Hussain Q, Usman A, et al. Impact of biochar properties on soil

conditions and agricultural sustainability: A review. Land Degradation and Development. 2017;**29**(7):2124-2161. DOI: 10.1002/ldr.2743

[44] Hossain MK, Strezov V, Chan KY, et al. Influence of pyrolysis temperature on production and nutrient properties of wastewater sludge biochar. Journal of Environmental Management. 2011;**92**(1):223-228. DOI: 10.1016/j.jenvman.2010.08.023

[45] Song W, Guo M. Quality variations of poultry litter biochar generated at different pyrolysis temperatures. Journal of Analytical and Applied Pyrolysis. 2012;**94**:138-145. DOI: 10.1016/j.jaap.2012.03.003

[46] Uchimiya M, Hiradate S. Pyrolysis temperature-dependent changes in dissolved phosphorus speciation of plant and manure biochars. Journal of Agricultural and Food Chemistry. 2014;**62**(8):1802-1809. DOI: 10.1021/jf405220k

[47] Sun H, Lu H, Chu L, et al. Biochar applied with appropriate rates can reduce N leaching, keep N retention and not increase NH_3 volatilization in a coastal saline soil. Science of the Total Environment. 2017;**575**:820-825. DOI: 10.1016/j.scitotenv.2016.09.173

[48] Wu LP, Zheng HN, Wang XJ. Effects of soil amendments on fractions and stability of soil organic matter in saline-alkaline paddy. Journal of Environmental Management. 2021;**294**:112993. DOI: 10.1016/j.jenvman.2021.112993

[49] Mandal S, Thangarajan R, Bolan NS, et al. Biochar-induced concomitant decrease in ammonia volatilization and increase in nitrogen use efficiency by wheat. Chemosphere. 2016;**142**:120-127. DOI: 10.1016/j.chemosphere.2015.08.008

[50] Liu L, Tan Z, Gong H, et al. Migration and transformation mechanisms of nutrient elements (N, P, K) within biochar in straw–biochar–soil–plant systems: A review. ACS Sustainable Chemistry and Engineering. 2019;**7**(1):22-32. DOI: 10.1021/acssuschemeng.8b05003

[51] Liang JF, Li QW, Gao JQ, et al. Biochar rhizosphere addition promoted Phragmites australis growth and changed soil properties in the yellow river delta. Science of the Total Environment. 2021;**761**:143291. DOI: 10.1016/j.scitotenv.2020.143291

[52] Weil R, Brady N. The Nature and Properties of Soils. 15th ed. Upper Saddle River: Prentice Hall; 2017

[53] Lashari MS, Ye Y, Ji H, et al. Biochar–manure compost in conjunction with pyroligneous solution alleviated salt stress and improved leaf bioactivity of maize in a saline soil from Central China: A 2-year field experiment. Journal of the Science of Food and Agriculture. 2015;**95**(6):1321-1327. DOI: 10.1002/jsfa.6833

[54] Zhang JY, Yang GJ, Yu YZ. Relationship between soil animal diversity and soil factors in Yinbei saline-alkali land. Northwest Agricultural Journal. 2012;**21**(7):177-184

[55] Manasa MRK, Katukuri NR, Darveekaran Nair SS, et al. Role of biochar and organic substrates in enhancing the functional characteristics and microbial community in a saline soil. Journal of Environmental Management. 2020;**269**:7. DOI: 10.1016/j.jenvman.2020.110677

[56] Shi S, Tian L, Nasir F, et al. Response of microbial communities and enzyme activities to amendments in saline-alkaline soils. Applied Soil

Ecology. 2019;**135**:16-24. DOI: 10.1016/j.apsoil.2018.10.003

[57] Li SP, Chen PZ, Zhou YY, et al. Effects of biochar application on available nutrients and enzyme activities in coastal saline-alkali soil. Journal of Southern Agriculture. 2019;**50**(7):1460-1465

[58] Zhang JY, Yang GJ, Ji CH. The relationship between the vertical distribution of soil arthropod and soil factors in saline-alkali Lycium barbarum plantation. Journal of Agricultural Sciences. 2016;**37**(3):11-17

[59] Liao S, Pan B, Li H, et al. Detecting free radicals in biochars and determining their ability to inhibit the germination and growth of corn, wheat and rice seedlings. Environmental Science and Technology. 2014;**48**(15):8581-8587. DOI: 10.1021/es501203k

[60] Buss W, Graham MC, Shepherd JG, et al. Risks and benefits of marginal biomass-derived biochars for plant growth. Science of the Total Environment. 2016;**569-570**:496-506. DOI: 10.1016/j.scitotenv.2016.05.173

[61] Rajapaksha AU, Chen SS, Tsang DC, et al. Engineered/designer biochar for contaminant removal/immobilization from soil and water: Potential and implication of biochar modification. Chemosphere. 2016;**148**:276-291. DOI: 10.1016/j.chemosphere.2015.12.052

[62] Chen H, Gao Y, Li J, et al. Engineered biochar for environmental decontamination in aquatic and soil systems: A review. Carbon Research. 2022;**1**(1):4. DOI: 10.1007/s44243-022-00004-3

[63] Trifunovic B, Gonzales HB, Ravi S, et al. Dynamic effects of biochar concentration and particle size on hydraulic properties of sand. Land Degradation and Development. 2018;**29**(4):884-893. DOI: 10.1002/ldr.2753

[64] Wang S, Gao B, Zimmerman AR, et al. Removal of arsenic by magnetic biochar prepared from pinewood and natural hematite. Bioresource Technology. 2015;**175**:391-395. DOI: 10.1016/j.biortech.2014.11.053

[65] Kambo HS, Dutta A. A comparative review of biochar and hydrochar in terms of production, physico-chemical properties and applications. Renewable and Sustainable Energy Reviews. 2015;**45**:359-378. DOI: 10.1016/j.rser.2015.01.010

[66] Wang J, Wang S. Preparation, modification and environmental application of biochar: A review. Journal of Cleaner Production. 2019;**227**:1002-1022. DOI: 10.1016/j.jclepro.2019.04.230

[67] Ullah I, Khan S, Nawab J, et al. Hardwood modified and unmodified biochar amendments used for saline alkali soil remediation: Phosphorus availability and its plant uptake. Arabian Journal of Geosciences. 2022;**15**(9):910. DOI: 10.1007/s12517-022-09000-0

[68] Duan ML, Liu GH, Zhou BB, et al. Effects of modified biochar on water and salt distribution and water-stable macro-aggregates in saline-alkaline soil. Journal of Soils and Sediments. 2021;**870**:2192-2202. DOI: 10.1007/s11368-021-03003-0

[69] Chen W, Zhao B, Guo Y, et al. Effect of hydrothermal pretreatment on pyrolyzed sludge biochars for tetracycline adsorption. Journal of Environmental Chemical Engineering. 2021;**9**(6):106557. DOI: 10.1016/j.jece.2021.106557

[70] Zhao SC, Yang JM, Zhou QH, et al. Effect of hydrothermal carbon amendment on forage growth and rhizosphere soil enzyme activities in saline-alkali soil. Journal of Jilin Agricultural University. 2024;**46**:860-868. DOI: 10.13327/j.jjlau.2021.1781

[71] Liu M, Linna C, Ma S, et al. Biochar combined with organic and inorganic fertilizers promoted the rapeseed nutrient uptake and improved the purple soil quality. Frontiers in Nutrition. 2022;**9**:997151. DOI: 10.3389/fnut.2022.997151

[72] Wen Liang W, Shi Rong Z, Li Peng W, et al. Biochar and phosphorus fertilization improved soil quality and inorganic phosphorus fractions in saline-alkaline soils. Archives of Agronomy and Soil Science. 2020;**67**(9):1177-1190. DOI: 10.1080/03650340.2020.1772303

[73] He XS, Zhang SQ, She D, et al. Effects of biochar on soil and fertilizer and future research. Chinese Agricultural Science Bulletin. 2011;**27**(15):16-25

[74] Lu H, Lashari MS, Liu X, et al. Changes in soil microbial community structure and enzyme activity with amendment of biochar-manure compost and pyroligneous solution in a saline soil from Central China. European Journal of Soil Biology. 2015;**70**:67-76. DOI: 10.1016/j.ejsobi.2015.04.003

[75] Prost K, Borchard N, Siemens J, et al. Biochar affected by composting with farmyard manure. Journal of Environmental Quality. 2013;**42**(1):164-172. DOI: 10.2134/jeq2012.0153

[76] Phuong NTK, Khoi CM, Ritz K, et al. Influence of rice husk biochar and compost amendments on salt contents and hydraulic properties of soil and rice yield in salt-affected fields. Agronomy. 2020;**10**(8):1101. DOI: 10.3390/agronomy10081101

[77] Xu G, Zhang Y, Sun J, et al. Negative interactive effects between biochar and phosphorus fertilization on phosphorus availability and plant yield in saline sodic soil. Science of the Total Environment. 2016;**568**:910-915. DOI: 10.1016/j.scitotenv.2016.05.173

[78] Liang S, Wang SN, Zhou LL, et al. Combination of biochar and functional bacteria drives the ecological improvement of saline-alkali soil. Plants. 2023;**12**(2):284. DOI: 10.3390/plants12020284

[79] Ndiate NI, Saeed Q, Haider FU, et al. Co-application of biochar and arbuscular mycorrhizal fungi improves salinity tolerance, growth and lipid metabolism of maize (*Zea mays* L.) in an alkaline soil. Plants (Basel). 2021;**10**(11):2490. DOI: 10.3390/plants10112490

[80] Li R, Zhang Y, Deng H, et al. Removing tetracycline and Hg(II) with ball-milled magnetic nanobiochar and its potential on polluted irrigation water reclamation. Journal of Hazardous Materials. 2020;**384**:121095. DOI: 10.1016/j.jhazmat.2019.121095

[81] Zhang P, Xue B, Jiao L, et al. Amelioration effects of coastal saline-alkali soil by ball-milled red phosphorus-loaded biochar. Chemical Engineering Journal. 2022;**431**:133904. DOI: 10.1016/j.cej.2021.133904

[82] Liu J, Jiang J, Meng Y, et al. Preparation, environmental application and prospect of biochar-supported metal nanoparticles: A review. Journal of Hazardous Materials. 2020;**388**:122026. DOI: 10.1016/j.jhazmat.2019.122026

[83] Xu X, Wang J, Tang Y, et al. Mitigating soil salinity stress with titanium gypsum and biochar composite materials: Improvement effects and mechanism. Chemosphere. 2023;**321**:138127. DOI: 10.1016/j.chemosphere.2023.138127

Chapter 6

Use of Biochar and Compost in Soil Remediation, Growth and Yield of Cereals and Legume Crops in Niger State, Nigeria

Adewale Nafiu, Victor Chude, Alex Coker and Godswill Aguiyi

Abstract

Iron toxicity has been one of the major limiting factors affecting crop production in Niger state. This occurrence is predominantly present in medium to strongly acidic soils of inceptisols, alfisols and ultisols soil classification taxonomy in the state. This constraint causes poor root and shoot development of the affected crops, leading to stunted growth, bronzing of the leaves and poor yield. This has greatly led to food insecurity and poor livelihood of smallholder farmers (SHFs) in the state. But, with the intervention use of biochar and compost on these iron toxicity soils in the 4 Local Government Areas (LGAs) of Niger state (Bosso, Gbako, Suleja and Wushishi), the plant height of the maize, rice and soybean crops was increased from farmers' practice (T1) 51.4–96.9, 21.4–25.8, 14.2–18.9 to 96.5–197.4, 38.2–78.5 and 25.3–50.1 cm, respectively, under the use of biochar, compost and inorganic fertilizer (T2); the number of leaves (NOL) increased from 14 to 16, 13–15, 10–16 to 22–24, 25–34 and 24–49, respectively, while there was also a better yield performance for rice from 2.0–2.30 Tons/Ha under farmer practice (T1) to 4.52–6.70 Tons/Ha under (T2) with the use of biochar, compost and inorganic fertilizer; maize from 1.42–1.84 Tons/Ha to 2.93–5.04 Tons/Ha, and soybean recorded value ranged from 0.66–0.95 Tons/Ha under T1 practice compared to 2.40–4.13 Tons/Ha recorded from T2 practice. The acidity of the soils was buffered from 5.7 to 6.3, fertility status of nitrogen, phosphorus, potassium and organic carbon and soil structures were also improved, thereby addressing the negative effects of iron toxicity on the crops.

Keywords: biochar, compost, organic fertilizers, Niger state, Nigeria

1. Introduction

One of the critical challenges limiting agricultural productivity in Niger state is the issue of soil degradation, arising from iron toxicity and making it a problem soil [1–4]. Problem soils are commonly defined as soils with undesirable physical, chemical and biological properties that inhibit or prevent plant growth [5]. This problem is prevalent in numerous soil classifications in the state such as inceptisols, alfisols and ultisols, especially acidic soils (lowland and upland), and is further worsened by the routine farm practices, including the overuse of chemical fertilizers, poor irrigation practices and externalities such as flooding [6]. Iron (Fe) toxicity is a plant nutritional problem associated with excess ferrous ion (Fe^{2+}), which is absorbed by the roots and transported by the xylem to the leaves, causing iron (Fe) overload in the plant tissue (**Figure 1**). The characterization of iron toxicity included low-pH dryness of the soil, high iron oxide content, poor drainage, low soil organic matter (OM), browning and drying of leaves among others, culminating in poor root development, plant growth, tillering and colossal yield reduction [7].

Studies have revealed that higher concentration of Fe^{2+} in the rhizosphere has antagonistic effects on the uptake of many essential nutrients [7]. While, low oxidation power of roots leads to more ferrous iron uptake that causes iron toxicity [8]. The yield loss induced by the iron toxicity is frequently associated with low nutritional status. Consequently, iron toxicity can be defined as a multiple nutritive disorder further increased by P, K and zinc (Zn) deficiencies through excesses of hydrogen sulfide (H_2S) [3, 7]. Symptoms of iron toxicity in rice result from an excessive uptake of ferrous ions by the roots and their acropetal translocation through xylem outflow to the aerial parts. The typical symptoms linked to this are the bronzing or yellowing processes of the leaves [9]. Yield losses associated with the occurrence of iron toxicity symptoms generally range from 15 to 50%. In the case of severe toxicity however, total harvest loss can occur (**Figure 2**) [10, 11].

Across Niger state, farmers have deployed several local mechanisms in addressing the recurring problem of iron toxicity in their soils. Key among these is the spreading of rice husks directly on the farmland; use of liming; use of raw organic matter, such as burnt rice husks at lower recommended temperature using drum; spreading of rice husks directly on the farmland and setting the farmland ablaze. However, these approaches have their limitations, including accessibility issues, partial control and repeated treatments, with obvious cost implications, prompt decomposition of materials, competition from other users, immobility and ultimately, unavailability of

Figure 1.
Bronzing of the soil is an indication of iron toxicity in Suleja LGA, Niger state.

DOI: http://dx.doi.org/10.5772/intechopen.1011738

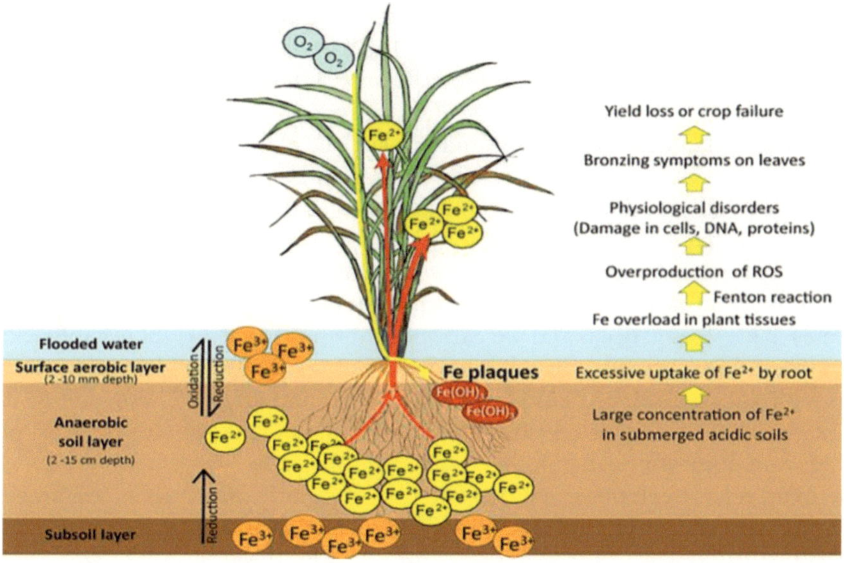

Figure 2.
Iron reduction in submerged low-pH soils and the effects of Fe toxicity on rice plants. Adapted from Vadugu et al. [10].

required nutrients in the soil, chemical pollution, increased global warming, among others. The deployment of burnt rice husks to produce black carbon with semblance to biochar is fraught with production-related challenges, arising from ignorance, limited capacity, huge environmental pollution and haphazard field applications due to the absence of soil testing (**Figure 3**) [5, 7]. This has not addressed the recurring challenge of iron toxicity in the state (**Figure 2**).

It is worthy of note that there are degradable dirts, garbage from the kitchen, domestic, industrial and agricultural wastes (animal and plant) found all over the environment because they were not properly disposed and taken care of. All these if not properly harnessed contribute to climate change, cause air, water and soil

Figure 3.
Burnt rice husks to produce black carbon with semblance to biochar by local farmer (L) and direct spread of rice husks on field (R) in Gbako LGA of Niger state.

pollution and directly affect many ecosystems, which result in poor agricultural yields and impacting negatively our environment [8]. The solution to this problem is the use of eco-friendly agricultural technology/innovation to transform these wastes to organic amendments for soil fertility/health improvement. This involves recycling process and recyclable biodegradable content such as plant-based materials into organic amendments. Also, reduction of polluting substances into landfill dump-like domestic waste, animal dung and agricultural wastes that may trigger greenhouse gas emissions into organic amendments.

Organic amendments occur naturally with or without the help of humankind. It's happening right now on forest floors, in farmers' fields and in our yards. But oftentimes, it's a slow process and we can speed it up using the right combination of water, oxygen, heat, organic materials and technology. There are fertilizers derived from organic sources, which could be of plant or animal origin. These include dead wood, plant debris, falling leaves and animal droppings (e.g., cattle manures, poultry droppings and domestic sewage).

Moreover, it has been observed that the use of organic materials as soil amendments improves environmental condition, public health as well as the quality of the crop. The fertility status of the soil is also improved by the activities of soil microorganisms [9, 11]. Consequently, He et al. [12] reported that the application of organic amendments plays vital roles in plant growth and also as source of all essential macro- and micro-nutrients in available forms during mineralization, thereby improving both physical and biological properties of the soil. This enhances soil productivity and increases crop yield.

Organic soil amendments like biochar and compost are widely suggested for the improvement of problematic soil due to their environmental friendliness, efficiency and economic viability [5]. The strategic addition of biochar and compost to problem soils has great potential to improve the quality of problem soils. Both biochar and compost are effective in alleviating the constraints of problem soils on plant growth or crop production, such as water scarcity, compaction, low fertility and pollutant toxicity [6].

In the last decade, it has been noticed that mixtures of biochar and compost can be applied to soils with excellent amelioration efficiency. Jien et al. suggested that compost and biochar together were more effective in stabilizing and sequestering carbon than biochar alone [6]. Meanwhile, the interaction between biochar and compost in combined applications has received growing attention [13, 14]. Biochar extends the residence time of compost in the soil, and compost solves the problem of insufficient nutrients provided by biochar [6, 15].

Studies have shown that biochar exhibits modifications after the co-composting process, such as surface functional groups, nutritional content, cation exchange capacity (CEC) and the amount of inhibitory aromatic compounds [13]. These changes have the potential to increase the efficiency of biochar and compost in soil improvement. For example, the increased nutrients and organic matter (OM) in biochar and compost may favor improved soil quality and microbial growth, which may benefit plant growth. The increased CEC of biochar and compost may be beneficial for improving nutrient retention capacity. An increase in the amount of functional groups in biochar and the number of microorganisms in compost may favor their ability to absorb pollutants [14]. Given these, biochar-compost has been considered as a potentially more effective soil amendment than biochar or compost.

Due to a lack of awareness and publicity about the effectiveness of biochar-compost on various problem soils, the willingness to utilize biochar-compost for soil improvement is still low compared to biochar or compost alone [13].

Because of all these properties mentioned above, this is why the study looked at the use of the biochar and amended compost in addressing the iron toxicity problem in the acidic soils of Niger state. Secondly to evaluate the growth and yield parameters of rice, maize and soybean in response to the use of biochar and compost in comparison to the farmers' practice in the state.

2. Materials and methods

2.1 Study area

The study was carried out in four (4) Local Government Areas (LGAs) of the state during the 2022 wet cropping season, namely: Bosso, Gbako, Suleja and Wushishi. The soils at the study area are classified mainly as inceptisols, alfisols and ultisols according to the U.S. Department of Agriculture (USDA) Soil Taxonomy system [16]. They are generally low in CEC, owing to the predominance of kaolinite in their clay fraction, low organic matter and inherent fertility. Three (3) commodity crops: rice, maize and soybean were used as test crops during the study.

2.2 Biochar production

Because of increasing interest of researchers in usage of biochar in the recent time for multiple purposes, the conversion of biomass, especially of agricultural wastes into biochar, is also growing. Biochar is found to have been produced with various methods, such as hydrothermal carbonization, gasification, torrefaction and flash carbonization and thermochemical conversion [11].

Biochar is usually produced through thermochemical conversion method. Thermochemical conversion processes include roasting, gasification, hydrothermal carbonization and pyrolysis [17, 18]. To obtain the highest possible yield of biochar, the production method must be adapted to the type of biomass and process parameters, such as heating rate, temperature and residence time. These conditions are critical because they affect the physical and chemical properties of biochar during the manufacturing process.

Since the process results in biomass loss, the form of biochar produced from plant biomass depends on process conditions. This study made use of the thermal conversion (pyrolysis) method to produce the biochar used for the treatment on the field.

Biochar, synthesized by pyrolyzing biomass, holds great chemical and biological stability, a high capacity for cation exchange, a large specific surface area and a great concentration of functional group [12]. Additionally, it aids in the improvement of growth of plants and microbial activity of soil, therefore leading to the acceleration of ecosystem restoring process [19].

The feedstock used for the biochar production was plant wastes (rice husks) obtained from rice milling centers across the 4 LGAs where the studies were conducted. This was chosen as our main feedstock because of its availability and accessibility across the locations. The feedstock (rice husks) were mopped up and prepared for use by feeding them into the biochar degrader and blending machine in the production center to reduce surface area. Afterward, a locally designed and fabricated biochar kiln was deployed for the production. The processing was done through pyrolysis method, which involved feeding the prepared feedstock (rice husks) into the biochar kiln through the inlet and heating the kiln in the absence of air to temperature between 400 and 450°C for at least 1 hour [20]. However, it is good to note that the accurate temperate is very important as a slight change in the temperature can render the product undesirable, therefore to ensure the accurate temperature was obtained and maintained to get the desirable product (biochar), a pyrometer was used to monitor the temperature in the kiln. The outlet of the heated kiln was opened to remove the hot product, which is the biochar, allowed it to cool by spreading to be air-dried and then bagged for use on the field (**Figure 4**).

2.3 Compost production

Innovative compositing technologies are evolving everyday, revolutionizing how we handle and utilize our agricultural and organic wastes. These technologies offer solutions for faster decomposition, improved nutrient content and reduced environmental impact such as vermicomposting, aerated static pile composting and in-vessel compositing. The latter method was employed in the compost production for this study.

In-vessel composting: It is a controlled environment, which allows for faster decomposition and better management of compositing process under anaerobic condition. First, a locally designed and fabricated compost bio-digester was developed using a sealed and closed metal object (see **Figure 5**).

Animal and plant wastes, such as dry cowdung, poultry droppings or dry grasses, were the materials used for the compost production. They are in abundance in the locations used for the studies, so it was easy to access and make available for the production. The production involved three stages mentioned below:

2.3.1 Feed preparation

The compost materials (cowdung, poultry waste, grasses or rice straw) were crushed to smaller sizes using the crusher/grinder.

2.3.2 Digestion stage

Afterward, the crushed compost materials mixture was fed into a locally designed and fabricated innovative compost bio-digester through the inlet and water was added to fill the two-third of the bio-digester, the remaining one-third was left to be occupied by biogas that was produced during the digestion stage of the compost production. The bio-digester opening was closed to prevent presence of air in the digestion process

Figure 4.
Locally designed and fabricated biochar kiln in use for the production of biochar at Ggbamanda production center in gidan mangoro village, Bosso LGA, Niger state.

Figure 5.
Locally designed and fabricated compost bio-digester used for compost production in the study.

(anaerobic), which is the best method World Health organization (WHO) recommends and also environmental friendly. Digestion of the compost materials took place for 21 days for initial production but subsequently, half of the digested materials were left in the bio-digester and fresh materials added to it, and it took 5 days for compete digestion to take place compared to 21 days that it took for the first digestion.

2.3.3 Disinfection stage

The digested materials were harvested from the bio-digester in the form of slurry compost, which contained beneficial and harmful microbes. There is need to cure the digested slurry compost, so it was subjected to heating at 100–110°C for 3 hours to disinfect it of the harmful microbes and remove the water content, then allow to cool, the assorted compost is ready for use on the field.

2.4 Biochar and compost characterization

The biochar and compost used for the study underwent characterization, including pH, and few elemental composition determinations. All the analyses were conducted using standard procedures.

3. Field studies

3.1 Plot size and experimental design

Field preparation was done mechanically with one plowing and harrowing. This was followed by manual ridging and mapping out of the plots. The plot size measuring 10 m × 5 m = 50 m^2 with 1 m alley in between plots was used. The plot size was manually prepared. The experimental design was randomized block design with three replications. The treatments consisted of T2: Half- recommended rate of NPK Tons/Ha + 2.5 Tons/Ha of biochar +2.5 Tons/Ha of compost; and T2—Farmer practices (full recommended rate of NPK Tons/Ha).

3.2 Sowing and spacing

Maize: An early maturity variety of maize (*Zea mays*) Samaz was planted on ridges, at an interrow spacing of 75 cm and intrarow distance of 25 cm. The seeds rate was two seeds per hole and later it thinned down to one plant per stand.

Rice: Faro 44 (improved variety) was planted on a flat seedbed, at a spacing of 20 cm × 20 cm and one seedling per stand.

Soybean: Soybean was planted at a spacing of 50 cm × 50 cm for an early variety sowed, then thinned to one stand at 3–4 weeks after planting (WAP).

3.3 Fertilizer application and data collection

Biochar and compost were directly applied and incorporated during the land preparation on the field at least 2 weeks before sowing/transplanting on plots T2 only for the compost to decompose and start releasing nutrients slowly into the soil and buffering of the acidity by the biochar, thereby reducing the iron toxicity in the soil. The inorganic fertilizer (NPK 20–10–10) was applied at 2 WAP using the deep placement method and then topdressed with urea in the form of urea super granule (USG) at 6 WAP with the same rate as during the first fertilizer application. All agronomical practices were duly observed. The agronomical parameters, such as plant height, number of leaves and yield, were measured at maturity. Data were subjected to two-way analyses of variance (ANOVAs) and means were separated with Least Significant Difference (LSD) at $P = 0.05$.

3.4 Soil testing

Before the study was conducted, the soil testing of the intervention areas was carried out to ascertain acidity, nutrients and fertility statuses. The ability of soil to sustain growth of agricultural crops, in terms of supply of essential nutrients, is referred to as soil fertility. Several parameters are considered in soil fertility assessment. For this study, soil pH, organic carbon (C), total nitrogen (N), available phosphorus (P), potassium (K) and iron (Fe) of the surface 20 cm soils were assessed to ascertain fertility, and Fe content and acidity status of the soils in the 4 LGAs where the study was conducted. Interpretation of results was based on fertility classification of the Soil Survey Staff and Chute et al. [16, 21].

4. Results and discussion

The analysis showed that the pH for biochar was 7.5 (alkaline), nitrogen, phosphorus and potassium contents of the biochar were 5.3, 2.0 and 1.8%, respectively, indicative that the nitrogen and potassium contents were high but that of phosphorus was

S/No	Samples analyzed	pH	OC (%)	N (%)	P_2O_5 (%)	K_2O (%)	Ca (%)	Mg (%)
1	Biochar	7.5	24.90	5.25	2.00	1.80	0.11	5.84
2	Compost	8.2	10.18	4.42	1.15	1.54	1.08	6.75

Table 1.
Shows the characterization of the amended compost and biochar produced through a locally designed and fabricated biochar kiln and bio-digester for this trial.

moderate, while the organic carbon contents stood at 42.9 and 24.9%, respectively, which were high as well. These results showed that the biochar is rich enough to buffer the acidity of the soil and improve soil structure. The same trend was obtained with compost analysis with pH values of 8.2 (alkaline), 4.42 (N), 1.15 (P) and 1.54 (K), respectively (**Table 1**).

The results of the soil testing of the 4 LGAs before the study was conducted showed that pH values ranged from 5.7 to 6.3, which reflect moderately to slightly acidic nature of the soils, thus making liming not necessary at that moment to manage the soil acidity problem. Organic matter plays an important role in holding soil nutrients and water. It also helps in binding mineral components to enhance soil structure. Organic carbon (C) content in the soils ranged from 6.06 to 8.09 g kg^{-1}, which implied that the soils have low reserves.

Iron (Fe) is a micro-nutrient and required by plants in minute quantity for normal growth. Although Fe is very active under low pH < 5.5, its high concentration in the soils of the 4 LGAs investigated suggests potential Fe-toxicity risk. While the essential nutrients (N, P, K) needed by the crops in large quantities are found to be very low in contents (0.18–0.25, 1.41–4.26, 0.01–0.61), respectively, thereby showing that the soils are acidic, have poor fertility status and Fe content is high. These soil chemical parameters were later investigated immediately at the end of this study and the results were remarkable, in the sense that the pH was buffered by the use of the biochar and the value has moved near to neutrality (6.3–6.8), organic carbon value increased to a range of 10.02–14.40 g kg^{-1}, while the essential nutrients (N, P, K) changed to 1.6–1.9, 5.12–7.04, 0.65–1.22, respectively, which is moderate from low initially [16, 22]. In conclusion, based on the findings of this soil testing, intervention in areas of increased crop production in these 4 LGAs through soil amendment (biochar and compost) used in combinations with mineral fertilizers proved the tendency to improve soil health and give bumper harvest in terms of productivity [21, 22].

4.1 Yield and growth parameters of the crops in response to the treatment levels during the study at LGAs in Niger state

4.1.1 Yield response of the crops to the treatment levels

The effect of treatments on the yield response of the three crops is presented in **Figure 6**.

4.1.1.1 Maize

Figure 6 shows the average yield range for maize in the 4 LGAs before the intervention as 1.42 to 1.84 Tons/Ha, which can be termed as low yield. But, after the intervention the yield ranged from 2.93 to 5.04 Tons/Ha under combined use of biochar, compost and inorganic fertilizers across the locations compared to the farmer practices. This is supported by the work of Jien et al. [15] that biochar application has positive effects on crop yield, soil properties and greenhouse gas emissions.

4.1.1.2 Rice

Yields ranged from 2.0 to 2.30 Tons/Ha under farmer practice (T1) to 4.52–6.70 Tons/Ha under (T2) use of biochar, compost and inorganic fertilizer. The yield advantage recorded is connected to the combined use of mineral fertilizers, organic

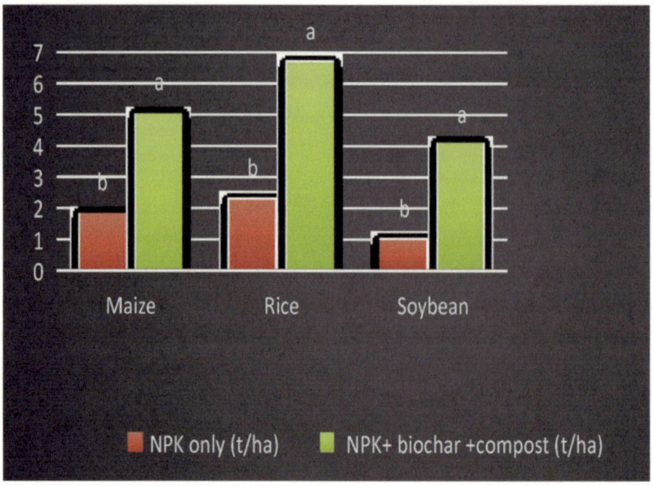

Figure 6.
The yield response of the crops (Tons/Ha) in response to the treatments in Niger state.

material and good agronomic practices, which enhanced agronomic use efficiency and crop productivity [17, 23] .

4.1.1.3 Soybean

For soybean, the yield range from T2 practice (2.40–4.13 Tons/Ha) across the 4 LGAs outweighed the yield range recorded from T1 practice (0.66–0.95 Tons/Ha). This is in line with previous results obtained from the other two crops (rice and maize).

4.1.2 Growth response to the treatment levels

The effect of treatments on the growth parameters (plant height and number of leaves/tillers) of the three crops is presented below in **Figures** 7 and **8**.

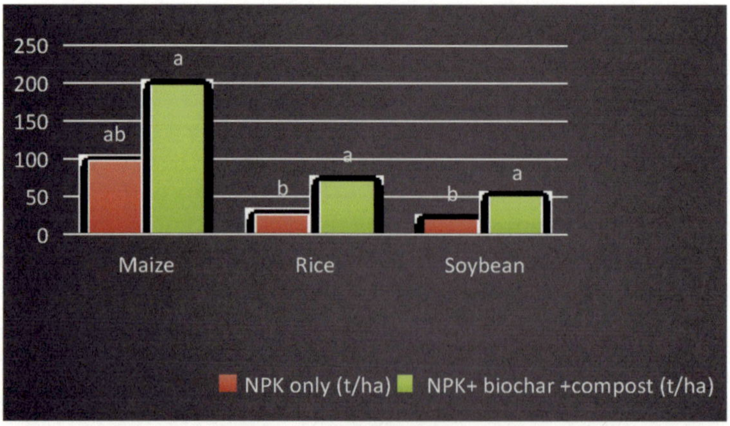

Figure 7.
The plant height (cm) of the crops in response to the treatments in Niger state.

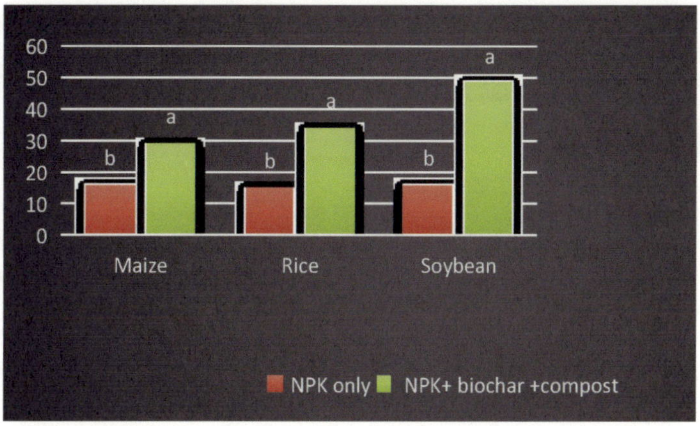

Figure 8.
The number of leaves (NOL) of the crops in response to the treatments in Niger state.

In **Figure 7**, the result showed that the plots treated with combined use of biochar, amended compost and inorganic fertilizer did better than the farmer practice with average plant height of the three crops (maize, rice and soybean) range values of 96.5–193.4, 38.2–78.5 and 25.3–50.1 cm, respectively, compared to the farmers' practice where only inorganic fertilizer was used. This result corroborated the study by the Soil Survey Staff and Pang [16, 18], where it was found that the application of organic and biofertilizers had positive effect on the root development and which resulted in more nutrient uptake and also promoted vigorous plant growth. This was also in line with the study conducted by Antonangelo et al. and Gour Das et al. [13, 24], which stated that, the increased nutrients and OM in biochar and compost favored improved soil quality and microbial growth, which in turn benefit plant growth.

In **Figure 8**, the number of leaves/tillers of the three crops (maize, rice and soybean) was greatly enhanced in the plots treated with combined use of biochar, compost and inorganic fertilizer with range values of 22–29, 25–34 and 24–49, respectively, compared to the farmers' practice where only inorganic fertilizer was used. This result corroborated with the earlier study carried out by Khan et al. and Arshad et al. [19, 25].

5. Conclusion

This study revealed that the application of biochar, amended compost and inorganic fertilizer to the plots where these three crops (maize, rice and soybean) were planted did better in terms of the highest plant growth and the number of leaves across the 4 LGAs in Niger state compared to the plot where NPK alone was applied, during the wet cropping seasons of 2022. There was also an upward total yield increase by 36.5% (maize), 34.5% (rice) and 23.2% (soybean) in plots treated with combined application of biochar, compost and NPK when compared to NPK-treated plot alone.

In addition to this, the soil acidity was buffered to near neutrality from 5.7 to 6.3 and the fertility status across the 4 LGAs was greatly improved with the combined application of biochar and amended compost during the land preparation as basal dressing, thus also addressing the issue of iron toxicity issue in the soils and increasing its productivity. Therefore, it is recommended that the use of biochar and compost in complementary with inorganic fertilizer should be adopted and upscaled to bring about effective soil fertility management and restoration mechanism in the identified problematic soils in Niger state and Nigeria at large.

Acknowledgements

I would like to acknowledge Alliance for Green Revolution in Africa (AGRA) for funding this study through their intervention project in Niger state, with the aim of building the resilience of smallholder farmers to soil degradation and climate change, improving their crop productivity and livelihood. Also, I acknowledged the management of National Program for Food Security (NPFS) under the Federal Ministry of Agriculture and Food Security for assigning me, the responsibility of implementing this project.

Conflict of interest

Authors have declared that no competing interests exist.

Thanks

Special thanks go to Niger State Ministry of Agriculture (MOA), Niger State Agricultural and Mechanization Development Agency (NAMDA) and other stakeholders that have seen this intervention work and impact positively the lives of SHFs in Niger state, Nigeria.

Acronyms and abbreviations

AGRA	alliance for green revolution in Africa
LGAs	local government areas
MOA	Niger state ministry of agriculture
NAMDA	Niger state agricultural and mechanization development agency
NPFS	National Programme for Food Security
USG	urea super granules
SHF	smallholder farmers
WHO	World Health Organization

Author details

Adewale Nafiu[1*], Victor Chude[1], Alex Coker[2] and Godswill Aguiyi[3]

1 National Programme for Food Security, Abuja, Nigeria

2 Department of Agric Economics, Federal University of Technology, Minna, Nigeria

3 Alliance for Green Revolution in Africa, Niger State Project Office, Minna, Nigeria

*Address all correspondence to: huskan2@gmail.com

References

[1] Kirk GJD. The Biogeochemistry of Submerged Soils. Chichester: U.K. John Wiley and sons; 2004. p. 291

[2] Audebert A, Sahrawat KL. Mechanism for iron toxicity tolerance in lowland rice. Journal of Plant Nutrition. 2000;**23**:1817-1885

[3] Olaleye O, Tabi FO, Ogunkunle A, Singh BN. Effect of toxic iron concentration on the growth of lowland rice. Journal of Plant Nutrition. 2007;**24**(3):441-457. DOI: 10.108/PLN-100104971

[4] Niger Consortium. Grant Proposal 'Improving Farmers' Resilience and Upscaling Productivity, Incomes and Livelihood in Selected Value Chains in Niger State. Niger Consortium; AGRA Publication; 2001. Available from: www.agra.org

[5] Yu H, Zou W, Chen J, Chen H, Yu Z, Huang J, et al. Biochar amendment improves crop production in problem soils: A review. Journal of Environmental Management. 2019;**232**:8-21

[6] Shixian Q, Xuerong Z, Yukui F, Biao S, Huchuan Y, Zhexin C, et al. Biochar-compost as a new option for soil improvement: Application in various problem soils. Science of the Total Environment. 2023;**870**(3). DOI: 10.1016/j.scitotenv.2023.162024

[7] Audebert A, Narteh LT, Kiepe P, Millar D, Beks B. Iron Toxicity in Rice-Based Systems in West Africa. Cotonou, Benin: Africa Rice Center (WARDA); 2006. p. 175

[8] Fagaria NK, Santos NAB, Barbosa MP, Quimanqes CM. Iron toxicity in lowland rice. Journal of Plant Nutrition. 2008;**31**(9):1676-1699

[9] Fagaria NK, Baligar VC. Lowland response to nitrogen fertilization. Communication in Soil Science and Plant Analysis. 2001;**32**(9):1405-1429

[10] Vadugu H, Shaik A, Shambhu C, Yuva P, Lalotra S. Iron toxicity in plants: A review. International Journal of Environment and Climate Change. 2023;**13**(8):1894-1900; Article no. IJECC.101675. ISSN: 2581-8627

[11] Becker M, Asch F. Iron toxicity in rice-conditions and management concept. Journal Plant Nutrition and Soil science. 2005;**168**(4):558-573

[12] He L et al. Remediation of heavy metal contaminated soils by biochar: Mechanisms, potential risks and applications in China. Environmental Pollution. 2019;**252**(Pt A):846-855

[13] Antonangelo JA, Sun X, Zang H. The roles of co-composted biochar (COMBI) in improving soil quality, crop productivity, and toxic metal amelioration. Journal of Environmental Management. 2021;**277**:pg 1-13. DOI: 10.1016/j.jenvman.2020.111443

[14] Wu S, He H, Inthapanya X, Yang C, Lu L, Zeng G, et al. Role of biochar on composting of organic wastes and remediation of contaminated soils—A review. Environmental Science and Pollution Research. 2017;**24**:16560-16577. DOI: 10.1007/s11356-017-9168-1

[15] Jien S-H, Chen W-C, Ok YS, Awad YM, Liao C-S. Short term biochar application induced variations in C and N mineralization in compost-amended tropical soil. Environmental

Science and Springer Publication. 2018;**25**:25715-25725

[16] Soil Survey Staff. Keys to Soil Taxonomy. 12th ed. Washington, DC: USDA - Natural Resources Conservation Service; 2014; 360 p

[17] Lin Y, Ma X, Peng X, Yu Z, Fang S, Lin Y, et al. Combustion, pyrolysis and char CO_2-gasification characteristics of hydrothermal carbonization solid fuel from municipal solid wastes. Fuel. 2016;**181**:905-915

[18] Pang S. Advances in thermochemical conversion of woody biomass to energy, fuels and chemicals. Biotechnology Advances. 2019;**37**:589-597

[19] Khan S, Irshad S, Mehmood K, Hasnain Z, Nawaz M, Rais A, et al. Biochar production and characteristics, its impacts on soil health, crop production, and yield enhancement: A review. Plants. 2024;**13**:166. DOI: 10.3390/ plants13020166

[20] Abou El-Magd MM, El-Bassiony AM, Fawzy ZF. Effect of organic manure with or without chemical fertilizers on growth, yield and quality of some varieties of broccoli plants. Journal of Applied Sciences Research. 2006;**2**:791-798

[21] Chude VO, Olayiwole SO, Asho AO, Daudu CK. Fertilizer Use and Management Practices for Crops in Nigeria. 4th Edition. Abuja: Federal Fertilizer Department, Federal Ministry of Agriculture and Rural Development; 2011. pp. 42-44

[22] Nafiu A, Yusuf AS, Jerry J, Yusuf S, Sharhabil MY. Biochar-enhanced soil pH and nutrient retention in Sudan savanna agricultural soils of Nigeria. Academic Environmental Sciences and Sustainability. 2005;**2**:1-8. DOI: 10.20935/ AcadEnvSci7664

[23] Li X, Wu D, Liu X, Huang Y, Cai A, Xu H, et al. A global dataset of biochar application effects on crop yield, soil properties, and greenhouse gas emissions. Scientific Data. 2005;**11**:57. DOI: 10.1038/s41597-023-02867-9

[24] Das Gour CM, Dharmendra P, Probir Kumar G, Shishomvanao N, Pankaj B. Productivity, nutrient uptake and postharvest soil fertility in lowland rice as influenced by composts made from locally available plant biomass. Archives of Agronomy and Soil Science. 2010;**56**(6):671-680

[25] Arshad M, Khalid A, Mahmood MH, Zahir ZA. Potentiality of nitrogen and L-tryptophan enriched compost for improving growth and yield of hybrid maize. Pakistan Journal of Agricultural Sciences. 2004;**41**:16-24

Chapter 7

Biochar as a Management Tool for Soilborne *Fusarium* and *Phytophthora* Pathogens

Kwasi Adusei-Fosu, Goldy De Bhowmick, Adnan Akhter and Muhammad Taqqi Abbas

Abstract

Biochar is a potential alternative management tool for diseases caused by soilborne pathogenic *Fusarium* and *Phytophthora*, respectively fungus and oomycete. Currently there are only general reviews for biochar applications against plant pathogens without emphasis on *Fusarium* and *Phytophthora*. Globally, soilborne *Fusarium* and *Phytophthora* are economically important pathogens that severely impact agriculture and forestry industries. Our present review focuses on biochar treatment against soilborne *Fusarium* and *Phytophthora*, biochar mechanism for *Fusarium* and *Phytophthora* suppression, and feedstock effect on biochar efficacy. Biochar applications alter the rhizosphere or soil environment which impedes the survival and reproduction of soilborne *Fusarium* and *Phytophthora* pathogens. Unlike other disease control approaches such as fungicides applications or fumigation that directly target the cell-wall and other structures in *Fusarium* and *Phytophthora* soilborne pathogens, biochar has different mechanisms of action. Over the years, majority of research have focused on biochar treatments against soilborne *Fusarium* or *Phytophthora* pathogens under controlled conditions but not field trials. While biochar has several advantages, evidence-based research is required to understand its impact on the environment. Future research areas for biochar includes evaluating applications for different farming systems, biochar efficacy ratings and validating application rates against *Fusarium* and *Phytophthora* soilborne pathogens.

Keywords: soilborne oomycete, soilborne fungus, plant health, efficacy, feedstock

1. Introduction

Soilborne pathogens are a huge problem for agriculture production [1] because they are hidden enemies of plants and offer greater challenge in their management. Among all known major plant diseases, soilborne pathogens alone are responsible for 90% of them [2]. Among soilborne pathogens, *Fusarium* and *Phytophthora* are considered as the most damaging plant diseases that threaten agriculture and forestry [3–5]. Globally, agriculture and forestry form the backbone of food security and economic growth [6, 7]. Therefore, it is important for *Fusarium* and *Phytophthora* soilborne

IntechOpen

pathogens to be managed. *Fusarium* species are well known to put economic burden or losses on the farming community [8] e.g., *Fusarium oxysporum* which is considered fifth most destructive persistent fungal pathogen [6]. Recent study reported the global impact of *Fusarium* spp. because the pathogen has a wide host-range presently in five major continents including Asia, Australia, North America, South America and Europe (Muhammad et al. 2023). *Phytophthora* is an emerging threat to the forest and horticulture sectors [9]. Soilborne *Phytophthora* species has been reported to cause 100% yield losses in some crops such as soybean [10].

Soilborne pathogens impact crop yields by infecting plants via their roots, causing wilting, discolouration and in some cases death of the plant. Soilborne pathogens reduce the value of agricultural lands because soilborne pathogens such as *Fusarium* can live in soil for over four decades [11–13]. This long-term effect of soilborne pathogens on the soil cause farmers to adopt land rotation farming system, but for locations where land is scarce, it becomes a daunting task. Over the years, chemicals (e.g., fumigation and fungicides) have been employed to control soilborne pathogens such as *Fusarium* and *Phytophthora* which are fungus and oomycetes, respectively [14, 15].

Chemicals are the main stay of farmers to manage diseases caused by *Fusarium* and *Phytophthora* spp. [16]. Different fungicides such as prochloraz, bromuconazole, benzamides, phosphonate, quinones are applied against *Fusarium* or *Phytophthora* soilborne pathogens based on their mode of actions and chemical properties [17]. However, an imperative drawback for fungicide treatment is the development of resistance in *Fusarium* and *Phytophthora* [18, 19]. Due to detrimental effects of chemicals on the environment and health of humans [20], there is the appetite for alternative tools to control soilborne diseases caused by *Fusarium* and *Phytophthora*.

Some 'natural enemies' or biological control agents such as *Trichoderma* spp., *Burkholderia cepacia*, *Pseudomonas* spp., plant growth-promoting rhizobacteria (PGPR) and arbuscular mycorrhizal fungi are also used to mitigate the losses caused by *Fusarium* and *Phytophthora* [21, 22]. However, the success of biological control depends not only on plant-microbial interactions, but also on the ecological fitness of the biological agents [23]. Identifying a soilborne disease control tool that would fine-tune soil physicochemical properties to suppress *Fusarium* and *Phytophthora* will be important for the agriculture industry. Biochar has been reported to modify soil physical, chemical, and biological properties to suppress soilborne pathogens [15, 24]. Therefore, application of soil organic amendments such as biochar serves as an alternative disease management option [25] against plant pathogens and diseases.

Biochar is a carbonaceous material derived from pyrolysis of biomass at temperatures 300–700°C under restricted oxygen supply [26]. The average carbon content of biochar lies between 70 and 80% and comprised of aromatic carbon structures including a crystalline phase and an amorphous phase. The crystalline phase is reformed as fused polyaromatic hydrocarbon sheets, and the amorphous phase is formed as random aromatic rings [26]. The carbon structure in biochar is a well-developed porous structure with various active surface functional groups, imparting multifunctional agricultural solutions most importantly, improving soil and plant health [27, 28]. Biochar is a known carbon-negative material [27], used for remediating polluted soil and water bodies, and has carbon reduction benefits [27, 29–34] with strong positive influence on climate change through limitation of emissions of greenhouse gases [30, 35, 36].

Generally, biochar is used as soil amendment and there have been several reports that demonstrated improvements in soil–plant health [35–39]. Biochar application provides favourable soil conditions such as pH [29, 40], water holding capacity,

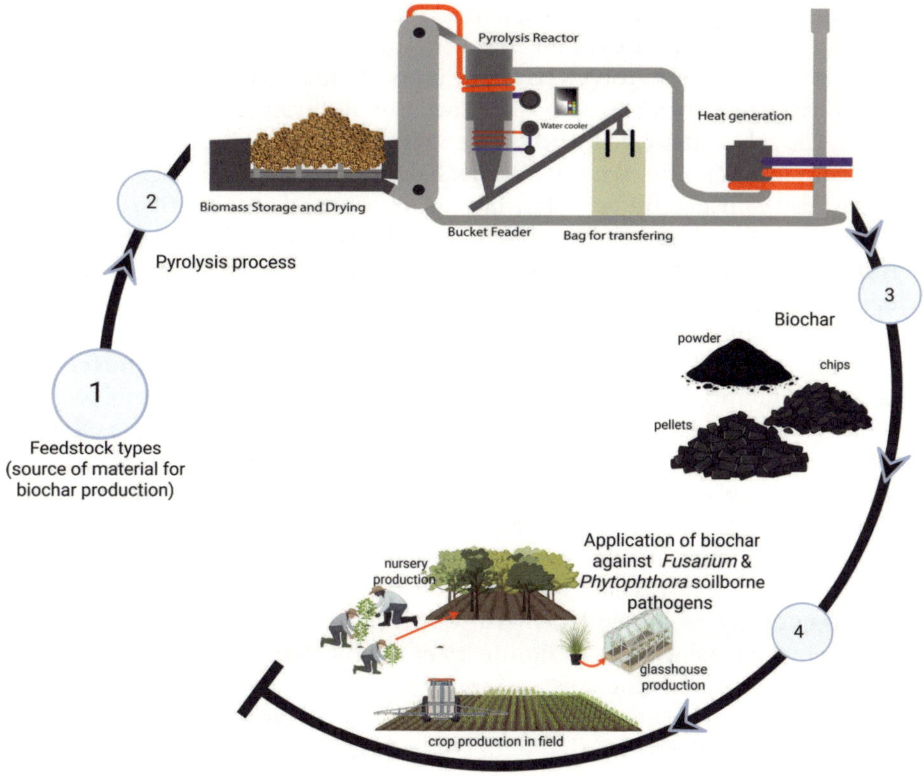

Figure 1.
Biochar production process flow chart and application.

nutrient supply (calcium-Ca, phosphorus-P, potassium-K, Nitrogen-N), cation exchange capacity (CEC) [29], the growth of beneficial bacteria [40] which enhances the overall soil and plant health. In addition, biochar does increase crop yield or biomass which has been demonstrated in maize [41], pepper [42], asparagus [43, 44], tomato [25], beans [45] and ryegrass [36].

In this review we have highlighted the application of biochar against *Fusarium* and *Phytophthora* soilborne pathogens from published literature (**Figure 1**). We have also discussed the effect of the pyrolysis process that influences the physicochemical composition biochar to enable disease management when used for soil treatments. Furthermore, we have highlighted the challenges associated with biochar applications under different agriculture production or farming systems.

2. Discussing the mechanisms of biochar in soilborne pathogen management

Thermochemical methods such as pyrolysis, gasification, hydrothermal liquefaction and torrefaction are used to convert organic matter of biomass into solid carbon residue biochar [46], but pyrolysis has gained more attention in the recent times because it yields byproducts including solid biochar [21, 46]. Previous studies have shown that pyrolysis temperature, duration, heating rate, vapour residence time, pyrolysis type, feedstock composition, and feedstock material/biomass properties including biomass type, moisture, and particle size play a pivotal role in influencing the physicochemical properties of biochar [21, 46].

Temperature used for pyrolysis has been reported to influence biochar properties with subsequent impact on soil. An increase in pyrolysis temperature increases the C, P, K, Ca, ash content, pH, and surface area of biochar, whiles decreasing N, H, and O content [47]. The structure of biochar obtained from high-temperature pyrolysis is distinguished by a substantial surface area and aromatic carbon content. On the other hand, soil treated with biochar produced via pyrolysis at low temperature yields high volatile and readily decomposable substrates which promote plant development [48]. The increase in pH is mostly triggered by breakdown of hydroxyl bonds present within the feedstock material under high pyrolysis temperature [49]. Such phenomenon also suggests the availability of weak bonds within the feedstock and their vulnerability towards thermal decomposition [49]. Generally, slow pyrolysis produces biochar with greater N, S, available P, Ca, Mg, surface area, and cation exchange capacity (CEC) in comparison to fast pyrolysis [47]. The surface area increases significantly due to the decomposition of organic matter, and destruction of aliphatic alkyls, esters, and the aromatic lignin core thereby reducing the pre-blocking substances and by increasing the externally accessible surface area [50, 51]. In some of the studies it was also observed that at temperatures lower than 500°C, the lignin is not fully converted to a hydrophobic polycyclic aromatic hydrocarbon (PAH) thereby increasing the hydrophilicity of the biochar [50, 51]. Thus, at temperatures higher than 650°C, biochar's are more hydrophobic and are thermally stable [50, 51]. The relationship between CEC and pyrolysis temperature was reported in previous studies and illustrated that with the increasing pyrolysis temperature CEC decreased linearly.

The physicochemical characteristic of biochar is determined by the initial feedstock and the pyrolysis conditions (heating temperature and time) [52]. Determining the best use for biochar requires comprehensive knowledge of the whole manufacturing process. Detailed physical, chemical characteristics and elemental composition of the biochar derived from different feedstocks have been researched by others [53]. The physicochemical characteristics of biochar have the potential to modify the availability of nutrients and carbon in the soil, as well as shield microorganisms from desiccation and predators. These changes might impact the soil's microbial diversity and taxonomy [38].

The pyrolysis temperature could yield different physicochemical properties of the biochar based on the feedstock type used. For instance, biochar's produced from woods, biosolids, and herbaceous feedstock demonstrated a neutral to slightly acidic pH upon lowering the pyrolysis temperature to under 400°C [49]. The variation in feedstock composition significantly affects the biochar properties including the elemental content (C, H, N, and S), yield, ash content, functional groups, aromaticity, porosity, specific surface area. According to [54], soil treated with biochar made from woodchip produced higher saturated hydraulic conductivities than those treated with manure-based biochar [55]. Similarly, manure-based biochar has a higher soil cation exchange capacity (CEC) than wood (*Eucalyptus*) biochar. To efficiently utilise majority of available biomass, the feedstocks are categorised into 1st to 3rd generation including lignocellulosic plants and energy crops, and algae. Recently, a fourth generation of biomass are being used widely includes organic waste such as coffee grounds, fruit peels, and sewage sludge [56–61]. Among the agricultural crop residues and woody biomass category corn stalks, wheat stalks, straw, rice husk, potatoes, soybeans, sugar cane, bagasse, cotton, oranges, grapes, peanuts, rapeseeds, etc. were explored for biochar production [56–61]. Previous studies revealed that biochar's made from agricultural wastes were mostly alkaline in pH under similar pyrolysis conditions.

In summary the effect of pyrolysis and feedstock types could enhance the biochar properties and subsequent positive impact on soil when used as a supplement. For instance, the CEC of biochar was associated with the disappearance of some acidic functional groups when pyrolysis temperature was increased [62]. Others have also reported that pyrolysis exhibited high porosity with longitudinal pores of sizes ranging from micro to macropores and the macropores were originated from vascular bundles of raw biomass that could provide habitats for symbiotic microorganisms promoting soil health [51]. The cellulose, and lignin present in the feedstock (crop residues) is reported to potentially decay into smaller fragments resulting in lower O/C and H/C [62]. However, in case of woody biomass the differences in hemicellulose, cellulose, and lignin proportions further affects the elemental composition with low ash, low moisture, high calorific value, and few voids.

2.1 Biochar treatments against soilborne pathogenic *Fusarium*

Fusarium is a cosmopolitan fungus and could be either soilborne or foliar pathogen [63]. They spread via the production of spores (conidia or microconidia), which act as the infecting materials as well as reproductive structures. These spores typically exist in either a haploid or dikaryotic vegetative stage [64]. Some soilborne *Fusarium* spp. are known to be host specific, whereas others have a broad host range. *Fusarium* research over the past 100 years has improved our understanding about the relevance of this fungi group, but there are still some unknowns regarding its biology that must be addressed [63]. Gaining better understanding of the biology will provide information to develop precise and reliable control tools against the *Fusarium* pathogen which are difficult to manage.

Soilborne *Fusarium* pathogens are very difficult to manage [65]. Others have resorted to control tools such as fumigation and fungicides treatments against soilborne *Fusarium* pathogens [14], but these methods are considered not environmentally friendly or sustainable. Chemicals usually directly target parts of the soilborne pathogens including cell-wall and other structures whiles contaminating the soil. However, developing sustainable methods that would not contaminate soil but modify its properties such as soil pH, nutrients supply/availability/mobility and beneficial microbial populations to inhibit pathogenic *Fusarium* spp. from thriving is an alternative sustainable method. For instance, it has been demonstrated that some soil amendments regulated soil pH and nutrient supply which contributed to the successful management of *Fusarium oxysporum* f. sp. *spinaciae* in spinach fields [14].

The application of biochar has been reported to suppress several pathogenic *Fusarium* spp. [25, 43, 44, 66–68]. A case study reported that biochar has been tested mostly against fungi soilborne pathogens including *F. oxysporum* f.sp. *radicis lycopersici*, followed by *F. oxysporum, F. solani* and *Fusarium* spp. [37, 69].

Earlier research shows that a commercial prototype biochar produced by fast pyrolysis of hardwood dust successfully reduced the damaging effect of allelopathy on arbuscular mycorrhizal (AM) root colonisation and on *Fusarium* crown and root rot of asparagus under greenhouse conditions [44]. Similar studies were done in greenhouses to determine biochar efficacy against *F. proliferatum* in infested soils collected from field and results showed that root lesions were reduced in the soil treated with biochar [43]. Others have demonstrated soilborne *F. oxysporum* f.sp. *lycopersici* chlamydospores were suppressed via amendment of soil with biochar produced from feedstocks including wood chip and green waste biochar [25, 66].

A rare experiment in field determined efficacy of biochar produced from poultry faecal waste and sawdust against pathogenic *F. verticillioides* causal organism for ear rot in maize, using biochar treatment combinations as soil amendments. Biochar from sawdust effectively managed the disease compared with poultry faecal waste [70]. Recent studies showed that biochar made from organic wastes for soil amendment significantly decreased the colonisation and survival of *F. oxysporum* f. sp. *vasinfectum* in potted cotton plant under controlled conditions [71].

Advances made in production of biochar has been the coupling of biochar with other microbial inoculants to provide multiple benefits to plants and soil. Most recently, several studies have shown that biochar produced from feedstock such as bone charcoal, wood chip mixed with garden waste residue or cassava, coupled with microbial inoculants including strains of *Pseudomonas chlororaphis, Paenibacillus polymyxa, Streptomyces pseudovenezuelae,* Arbuscular mycorrhizal fungi, and *Trichoderma aureoviride*, suppressed pathogenic *F. solani, F. oxysporum* f.sp. *radicis lycopersici* in soil [72]. Identifying an approach to develop appropriate coupling materials to improve the biochar efficacy against *Fusarium* soilborne is very important for plant protection.

2.2 Biochar treatments against soilborne pathogenic *Phytophthora*

Phytophthora is an oomycete and could be soilborne or foliar pathogen. They mainly spread via the movement of infested soil, water, or plants using their two flagella that enhance swimming ability of zoospores [73]. The vegetative stage of oomycetes is diploid [64] and have cellulose in its cell walls [74]. *Phytophthoras* are known to have a wide host range [75], however, there are other host specific species such as the *Phytophthora agathadicida* [76] among others. Most agricultural practices rely heavily on chemicals for the management of *Phytophthora* soilborne pathogens. Applied chemicals target the cell walls of these pathogens but due to the difference in oomycete cell wall structure, a different strategy is required to control oomycetes diseases [77]. Thus, if targeting the pathogen directly has existing challenges, then other mechanisms of controlling phytophthora species via altering the soil environment would be a promising method such as biochar treatments.

There is lack of information on the use of biochar to treat *Phytophthora* pathogens. A few studies have reported its effectiveness in controlling several *Phytophthora* soilborne pathogens including *P. cinnamomi*, *P. capsici, P. cactorum* via soil amendment with biochar [42, 78, 79]. Earlier research explored the effect of biochar on *P. cinnamomi* and *cactorum* and suggested that biochar amendment has the potential to alleviate disease progression and physiological stress caused by phytophthora canker pathogens [78]. In addition, they reported that there is likely an optimal level of biochar incorporation into the root media, beyond which its beneficial effects may be less pronounced [78].

Other trials have been conducted to determine whether two commercially available biochar amendments could suppress *P. capsici* infection in sweet bell pepper (*Capsicum annuum*) using three naturally infested field soils. Their research showed that amending soil with a biochar product containing a proprietary mix of beneficial microorganisms and enriched substrates resulted in a lower soil *P. capsici* abundance in all soils, and low root infection in two of the three soils tested. This treatment also led to increased soil pH, lower soil nitrogen availability, and reduced leaf chlorophyll content [79]. Recent research has demonstrated that biochar treatments effectively inhibited pathogen growth and controlled *P. capsici*, with biochar applied immediately before planting having greater effects than that applied 20 days before planting [42]. Further trials compared the control treatments with biochar-amended rhizosphere soils and reported that biochar treated soil had higher pH levels, more available nutrients (e.g., P and K) and higher abundance of beneficial fungi that led to the suppression of *P. capsici* [42]. There is less information for field application of biochar against pathogenic *Phytophthora*. As it stands all the research with biochar treatments against *Phytophthora* have mainly been conducted under controlled conditions [42, 69, 78, 79]. It is also evident that most of the commercial and newly developed biochar products have not been tested against phytophthora on a commercial scale. This is concerning because farmers need to rely on research data to make informed decisions regarding the use of biochar products against *Phytophthora* soilborne pathogens. In a recent case study, it was reported that biochar has been tested against only 6% of oomycetes [69], highlighting the lack of research with biochar against pathogenic oomycetes. Interestingly, at the time only one oomycetes *P. capsici* was identified as soilborne with the rest foliar pathogens [69].

2.3 Biochar mechanisms of action against *Fusarium* and *Phytophthora* soilborne pathogens

Biochar can alter soil physicochemical properties such as soil pH, water holding capacity, bulk density, soil aggregation [35], which subsequently enhance the soil to suppress soilborne pathogens as shown in **Figure 2** and **Table 1**. However, the effectiveness of biochar to alter any of these soil properties to suppress soilborne diseases depends on the application rate [24]. A major property of soil that leads to the control of soilborne pathogens is soil pH [14, 24, 79, 81–84]. There have been several reports that showed that changes in soil pH can successfully control soilborne *Fusarium* wilt diseases [14, 24, 35, 81–84]. Similarly, an oomycete soilborne pathogen such as *Phytophthora* has been reported to be suppressed by changes in soil pH, and this has also been reported for *Phytophthora* blight disease [79] including other *Phytophthora* soilborne pathogens. In vitro studies have shown that acidification suppresses the natural capacity of soil microbiome such as bacterial communities to fight pathogenic soilborne *Fusarium* infections because soil acidification disrupts the emission of antifungal substances [72]. The role of pH has been reported to transform soil via contributing atmospheric sulphur dioxide and nitrogen oxides because of the effect on bacterial populations in soil [85, 86]. Research has shown that sulphur compounds severely inhibits the growth of pathogenic fungi through the destruction of cell membranes or disruption of cellular antioxidation systems [86]. Under in vitro conditions, the optimal pH at which pathogen survived differed with among *Phytophthora* pathogens. For instance, an in vitro experiment showed that the optimum pH for *P. citricola* to survive was at 9, whereas *P. tropicalis* at pH 5, and five other species, *P. citrophthora*, *P. insolita*, *P. irrigata*, *P. megasperma*, and *P. nicotianae*, at pH 7 [87]. In another study, biochar was used to also suppress *Phytophthora* spp. in pineapple fields by amending soil with sulphur to lower pH from 6.3–5.1/5.4 [88]. Others have attributed the biochar induced inhibition of *Phytophthora capsici* to the augmented action of beneficial fungal microbial community of the soil [42]. Biochar induced resistance mechanisms were reported in *Quercus rubra* and *Acer rubrum* against *P. cinnamomi*, *P. cactorum*, respectively [78].

Nutrient availability is another indicator that regulates the dynamics of soilborne pathogen and plant health [79]. Biochar amendment is reported to contribute to soil

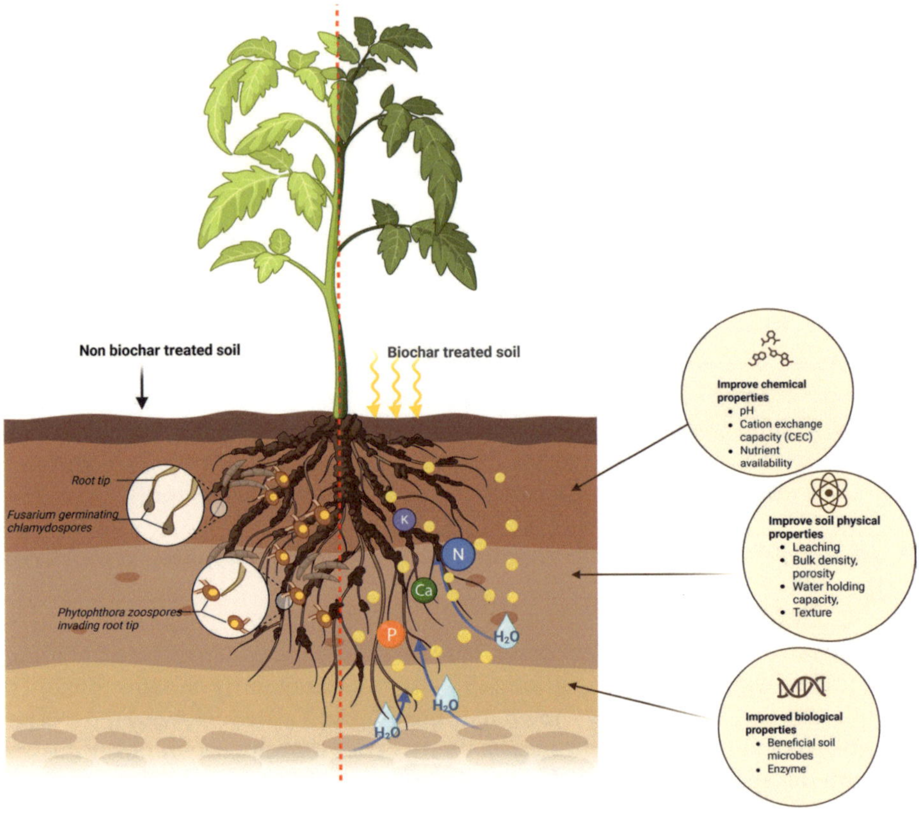

Figure 2.
Picture showing the effect of biochar on soilborne Fusarium *and* Phytophthora *pathogens.*

nutrition through the release of available nutrients to the roots of plants which leads to the decrease in root hairs development [39], thereby reducing the surface area for root infection by soilborne pathogens [89]. A study showed that considerable amounts of P, K, and Ca with limited amount of Mg contributed to a reduction of soilborne pathogens such as *Phytophthora* [79] and *Fusarium* [90].

Biochar has variable amount of anti-oxidants, phenolics and other organic acids [91]. These compounds can be toxic at higher concentrations to activates plant defence systems [66, 92]. Biochar-borne organic stimuli showcase the potential to both activate both salicylic acid (SA) and jasmonic acid (JA), pathways [92–94]. Moreover, biochar can absorb toxins and accelerate the production of antibiotics together with influence on the chemical composition of root exudation, collectively inhibiting establishment of soilborne pathogens on plants. As a result of hermetic response of biochar borne chemicals, augmentation of resistance pathways has been extensively reported [66, 95].

The successful disease infection process by a pathogen relies on environmental factors (e.g., humidity, temperature, rainfall), susceptible host plant and the pathogen as depicted in the disease triangle [96]. This means that for biochar to successfully influence the soil physicochemical properties to suppress soilborne pathogens, it would need a favourable environmental and a mechanism to trigger some defences in the host plant against the pathogens. Although some studies have explored how biochar changes the dynamics of the soil physicochemical properties to suppress

Pathogen class	Pathogen name	Host	Treatments (Biochar or supplemented with other agents)	Reference
Fungi	*Fusarium virguliforme*	Soybean	Biochar	[69]
	Fusarium solani	Cassava	Biochar coupled with microbial inoculant	[72]
	Fusarium oxysporum f.sp. radicis lycopersici	Tomato	Biochar coupled with microbial inoculant	[72]
	Fusarium oxysporum f. sp. *vasinfectum*	Cotton	Biochar	[71]
	Fusarium oxysporum f.sp. *lycopercisi* *Fusarium oxysporum* f.sp. *asparagi*	Tomato Asparagus	Biochar Biochar	[24, 44]
	Fusarium oxysporum f.sp. *radicis lycopersici*	Tomato	Biochar combined with phosphate-mobilising bacteria	[80]
	Fusarium oxysporum f.sp. *lycopersici*	Tomato	Biochar combined with either compost, Arbuscular mycorrhizal fungi	[25, 66]
	Fusarium spp.	Not available		[68]
	Fusarium proliferatum	Asparagus	Biochar	[43]
	Fusarium oxysporum f.sp. *asparagi*	Asparagus	Biochar	[43]
	Fusarium solani	Asian ginseng	Biochar	[67]
	Fusarium verticilloides	Maize	Biochar	[70]
	Fusarium virguliforme	Soybean	Biochar	[69]
	Fusarium oxysporum f. sp. *vasinfectum*	Cotton	Biochar	[71]
Oomycetes	*Phytophthora capsici* L. *Phytophthora infestans* (Mont.) de Bary *Phytophthora infestans* (Mont.) de Bary	Pepper Tomato Potato	Biochar combined with beneficial microbes Biochar coupled with microbial inoculants Biochar combined with *Streptomyces* strains	[24, 42, 60]
	Phytophthora nicotianae	Tobacco	Biochar coupled with microbial inoculants	[72]
	Phytophthora cinnamomi	Northern red oak	Biochar	[78]
	Phytophthora cactorum	Red maple	Biochar	[78]
	Phytophthora capsici L.	Sweet bell pepper	Biochar mixed with beneficial microorganisms	[79]

Table 1.
Literature confirming Fusarium *and* Phytophthora *soilborne pathogens treated with biochar as a stand-alone or supplemented with other disease control agents.*

soilborne pathogens including *Fusarium* and *Phytophthora*, the effect of the environmental factors on applied biochar, and the mechanisms that could trigger defences in the host against soilborne pathogens remains a conundrum.

3. Limitations for biochar treatments against soilborne *Fusarium* and *Phytophthora*

The optimal dosage of biochar application is still not established because it depends on biochar type, soil types and target species. Earlier research elucidated that an application rate of biochar 0.5 to 135 t ha^{-1} supported positive plant growth responses [97]. Whereas, others have reported variable response of biochar type and application rate on the pathosystems [35]. Recent reports suggest inconsistent results for biochar tested at different concentrations against soilborne pathogens. For instance, *Ralstonia solanacearum* development in eggplants was suppressed at a lower concertation of 6% leaf waste biochar [98]. However, biochar of pinewood origin had more suppressive effective at a higher (20%) concentration against peach replant disease [99]. A deeper understanding of the mechanisms supporting biochar-mediated disease suppression is important for the development of products that can consistently reduce soilborne disease in the field [79]. Unfortunately to date, the scientific community lacks information for standardised biochar application rates and feedstock types used against soilborne *Fusarium* and *Phytophthora*. Acquiring data for biochar application rates, feedstock types and pyrolysis processes to support production of commercial biochar will massively transform the biochar and agriculture industries [60]. An approach for standardising biochar application rates will depend heavily on an established databank for biochar results showing specific application rates juxtaposed with the feedstock type, source, age, and whether biochar is made from standalone organic source or mixture. Information for soil type, and stage of soilborne *Fusarium* and *Phytophthora* pathogen treated will be relevant for adoption.

4. Prospects of biochar: Operationalising biochar for agricultural production systems

Although biochar is currently being advocated to be integrated into farming practices (**Figure 3**) as a soilborne disease or pathogen control tool, less is communicated about the most ideal application techniques. After testing the efficacy of biochar against soilborne diseases caused by *Fusarium*, *Phytophthora*, or other pathogens, the goal is to adopt and implement the findings to suit various farming systems. Current research has demonstrated that biochar is a promising soil amendment that controls soilborne *Fusarium* and *Phytophthora* pathogens [24, 25, 35, 43, 44, 67–70, 78, 79, 84]. Majority of research confirming efficacy of biochar against *Fusarium* and *Phytophthora* soilborne pathogens have been experimented in potted plants under greenhouse or glasshouse conditions and this presents an opportunity for biochar application in nurseries where plant or crops are produced.

Other nurseries that use bare-root production systems to produce planting materials may contribute to the spread of soilborne pathogens such as *Fusarium* and *Phytophthora* [100–103]. Therefore, integrating biochar treatments into the nursery for plant production system as a disease control tool could serve as biosecurity tool to ensure clean healthy plant materials are supplied to farmers. This would limit the risk of introducing and spreading soilborne *Fusarium* or *Phytophthora*.

Greenhouse production systems could also benefit immensely from biochar as a soilborne disease control option. Vegetable crops such as tomato and pepper, which are also impacted by *Fusarium* and *Phytophthora* soilborne pathogens [25, 42, 79], are produced in both small- or large-scale commercial greenhouses. In colder countries

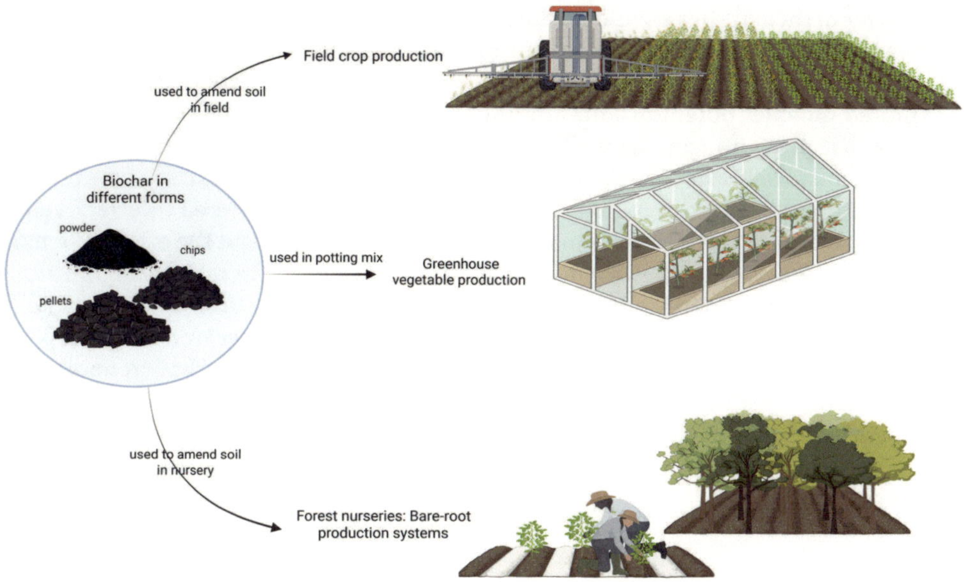

Figure 3.
Application of biochar in different agricultural production systems to manage soilborne Fusarium *and* Phytophthora.

such as Canada, the commercial greenhouse production systems are used for vegetable cultivation throughout the year. This is also true for other countries where farmlands are scarce. Treatment of soil or potting mix with biochar for vegetable production under greenhouse conditions could be integrated into the production practices to limit chemical treatments to manage pathogens such as *Phytophthora capsici* L. and *F. oxysporum* f.sp. *lycopercisi* for pepper and tomato respectively [24, 42, 79].

Although biochar has proved to be effective against soilborne pathogens *Fusarium* and *Phytophthora*, its large-scale application on farms may face economic challenges. More research is needed to explore the best biochar application methods for larger plantations where soils have been severely infested with *Fusarium* and *Phytophthora* pathogens. A comprehensive cost–benefit analysis is required for farmers operating on large farmlands who would want to explore the biochar treatments taking into consideration the direct and indirect cost in comparison with any existing *Fusarium* and *Phytophthora* soilborne disease management tools they are using.

5. Conclusions

Biochar as a soil amendment plays a crucial role in suppressing soilborne *Fusarium* and *Phytophthora* pathogens. However, there is still a significant knowledge gap regarding the application rates and extensive field trials for existing commercial biochar products from different feedstocks. It is imperative for more research to be conducted to establish appropriate application guidelines for farmers to make informed decisions prior to using biochar as soilborne diseases management tool. Although soil treatment with biochar could result in disease suppression, there is a risk that disease incidence could still exceed the economic threshold. Thus, attention should be given to biochar treatments that could suppress diseases from reaching this economic threshold. Further studies should be done to explore how environmental factors and plant host defence triggered mechanisms interact with biochar-treated soil against soilborne *Fusarium* and *Phytophthora* pathogens. Other aspects of biochar that must be explored includes potential confounding effects on humans and the environment to ensure that pros and cons are well understood. Most reported research on biochar efficacy against soilborne *Fusarium* and *Phytophthora* pathogens has been conducted under controlled conditions in pots. However, few studies have tested its effectiveness in the field. The lack of field trials makes it difficult to understand the effect biochar could have on *Fusarium* and *Phytophthora* soilborne pathogens in real field conditions. Another important reason why field trial is required to validate biochar efficacy is because of the heterogeneity nature of soil unlike potted soil. Research to develop biochar would be successful provided bioprocessing engineers, biochar technologists, and plant pathologists establish meaningful collaborations to explore the full potential of biochar for controlling soilborne diseases caused by *Fusarium* and *Phytophthora* pathogens. Involving policy advisors is also paramount to guide the adoption of biochar applications.

Acknowledgements

We appreciate the support of all the contributors of this review who committed their time to review material prior to submission. We also thank the editorial team that provided constructive feedback. We thank Prof. Hailong Wang at Foshan University, China and the Director of The Biochar Engineering Technology Research Centre of Guangdong Province for reviewing the manuscript and making constructive suggestions. We also thank Dr. Jiafa Lou for his constructive review.

Declaration of competing interest

The authors declare that they have no known competing financial interests or personal relationships that could have appeared to influence the work reported in this paper.

Author details

Kwasi Adusei-Fosu[1]*, Goldy De Bhowmick[2], Adnan Akhter[3]
and Muhammad Taqqi Abbas[3]

1 Resilient Agriculture, AgResearch Ltd., Grasslands Research Centre, Palmerston North, New Zealand

2 Smart Foods and Bioproducts, AgResearch Ltd., Te Ohu Rangahau Kai, Palmerston North, New Zealand

3 Faculty of Agriculture Sciences, Department of Plant Pathology, Quaid-e-Azam Campus, University of the Punjab, Lahore, Pakistan

*Address all correspondence to: kwasi.adusei-fosu@agresearch.co.nz

References

[1] Shafique HA, Sultana V, Ehteshamul-Haque S, Athar M. Management of soil-borne diseases of organic vegetables. Journal of Plant Protection Research. 2016;**56**:222-230

[2] Mokhtar M, El-Mougy N. Bio-compost application for controlling soil-borne plant pathogens–a review. International Journal of Engineering and Innovative Technology. 2014;**4**:61-68

[3] Bose T, Spies CF, Hammerbacher A, Coutinho TA. Phytophthora: An underestimated threat to agriculture, forestry, and natural ecosystems in sub-Saharan Africa. Mycological Progress. 2023;**22**:78

[4] Hardy GSJ. Phytophthora root rot of Forest trees. In: Encyclopedia of Forest Sciences. Academic Press; 2004. pp. 758-766

[5] Yan X, Guo S, Gao K, Sun S, Yin C, Tian Y. The impact of the soil survival of the pathogen of *Fusarium* wilt on soil nutrient cycling mediated by microorganisms. Microorganisms. 2023;**11**:2207

[6] Dean R, Van Kan JA, Pretorius ZA, Hammond-Kosack KE, Di Pietro A, Spanu PD, et al. The top 10 fungal pathogens in molecular plant pathology. Molecular Plant Pathology. 2012;**13**:414-430

[7] Jhariya MK, Banerjee A, Meena RS, Yadav DK. Agriculture, forestry and environmental sustainability: A way forward. In: Sustainable agriculture, forest and environmental management. Singapore: Springer; 2019. pp. 1-29

[8] Bockus WW, Bowden RL, Hunger RM, Murray T, Smiley R. Compendium of Wheat Diseases and Pests. USA: American Phytopathological Society (APS Press); 2010

[9] Hansen EM. *Phytophthora* species emerging as pathogens of forest trees. Current Forestry Reports. 2015;**1**:16-24

[10] Dorrance AE. Management of *Phytophthora sojae* of soybean: A review and future perspectives. Canadian Journal of Plant Pathology. 2018;**40**:210-219

[11] Adusei-Fosu K. Improving the Detection and Control of Fusarium Oxysporum f. Sp. Elaeidis. UK: University of Nottingham; 2017

[12] Paugh KR, Gordon TR. Survival of *Fusarium oxysporum* f. sp. lactucae on crop residue in soil. Plant Disease. 2021;**105**:912-918

[13] Scott J, McRoberts D, Gordon T. Colonization of lettuce cultivars and rotation crops by *Fusarium oxysporum* f. sp. lactucae, the cause of *Fusarium* wilt of lettuce. Plant Pathology. 2014;**63**:548-553

[14] McDonald MR, Collins B, duToit L, Adusei-Fosu K. Soil amendments and fumigation for the management of *Fusarium* wilt of bunching spinach in Ontario, Canada. Crop Protection. 2021;**145**:105646

[15] Parajuli M, Baysal-Gurel F. Control of *Phytophthora* and *Rhizoctonia* root rot on red maple using fungicides and biofungicides. HortScience. 2022;**57**:1306-1312

[16] Foster J, Hausbeck M. Managing *Phytophthora* crown and root rot in bell pepper using fungicides and host resistance. Plant Disease. 2010;**94**:697-702

[17] Amini J, Sidovich D. The effects of fungicides on *Fusarium* oxysporum f. sp. lycopersici associated with *Fusarium* wilt of tomato. Journal of Plant Protection Research. 2010;**50**:172-178

[18] Hu C-H, Perez FG, Donahoo R, McLeod A, Myers K, Ivors K, et al. Recent genotypes of *Phytophthora infestans* in the eastern United States reveal clonal populations and reappearance of mefenoxam sensitivity. Plant Disease. 2012;**96**:1323-1330

[19] Patel JS, Vitoreli A, Palmateer AJ, El-Sayed A, Norman DJ, Goss EM, et al. Characterization of *Phytophthora* spp. isolated from ornamental plants in Florida. Plant Disease. 2016;**100**:500-509

[20] Harsimran Kaur G, Harsh G. Pesticides: Environmental impacts and management strategies. In: Marcelo LL, Sonia S, editors. Pesticides. Rijeka: IntechOpen; 2014. Ch. 8

[21] Das MM, Haridas M, Sabu A. Biological control of black pepper and ginger pathogens, *Fusarium oxysporum*, Rhizoctonia solani and *Phytophthora capsici*, using Trichoderma spp. Biocatalysis and Agricultural Biotechnology. 2019;**17**:177-183

[22] Larkin RP, Fravel DR. Efficacy of various fungal and bacterial biocontrol organisms for control of *Fusarium* wilt of tomato. Plant Disease. 1998;**82**:1022-1028

[23] Alabouvette C, Olivain C, Migheli Q, Steinberg C. Microbiological control of soil-borne phytopathogenic fungi with special emphasis on wilt-inducing *Fusarium oxysporum*. The New Phytologist. 2009;**184**:529-544

[24] Jin X, Zhou X, Wu F, Xiang W, Pan K. Biochar amendment suppressed *Fusarium* wilt and altered the rhizosphere microbial composition of tomatoes. Agronomy. 2023;**13**:1811

[25] Akhter A, Hage-Ahmed K, Soja G, Steinkellner S. Potential of Fusarium wilt-inducing chlamydospores, in vitro behaviour in root exudates and physiology of tomato in biochar and compost amended soil. Plant and Soil. 2016;**406**:425-440

[26] Saravanan A, Kumar PS. Biochar derived carbonaceous material for various environmental applications: Systematic review. Environmental Research. 2022;**214**:113857

[27] Wu P, Wang Z, Bolan NS, Wang H, Wang Y, Chen W. Visualizing the development trend and research frontiers of biochar in 2020: A scientometric perspective. Biochar. 2021;**3**:419-436

[28] Yadav SPS, Bhandari S, Bhatta D, Poudel A, Bhattarai S, Yadav P, et al. Biochar application: A sustainable approach to improve soil health. Journal of Agriculture and Food Research. 2023;**11**:100498

[29] Gao Y, Chen H, Fang Z, Niazi NK, Adusei-Fosu K, Li J, et al. Coupled sorptive and oxidative antimony(III) removal by iron-modified biochar: Mechanisms of electron-donating capacity and reactive Fe species. Environmental Pollution. 2023;**337**:122637

[30] Hussain MM, Bibi I, Ali F, Saqib ZA, Shahid M, Niazi NK, et al. The role of various ameliorants on geochemical arsenic distribution and CO_2-carbon efflux under paddy soil conditions. Environmental Geochemistry and Health. 2023;**45**:507-523

[31] Liu J, Wei X, Ren S, Qi J, Cao J, Wang J, et al. Synergetic removal of

thallium and antimony from wastewater with jacobsite-biochar-persulfate system. Environmental Pollution. 2022;**304**:119196

[32] Luo J, Li X, Ge C, Müller K, Yu H, Deng H, et al. Preparation of ammonium-modified cassava waste-derived biochar and its evaluation for synergistic adsorption of ternary antibiotics from aqueous solution. Journal of Environmental Management. 2021;**298**:113530

[33] Niazi NK, Bibi I, Shahid M, Ok YS, Shaheen SM, Rinklebe J, et al. Arsenic removal by Japanese oak wood biochar in aqueous solutions and well water: Investigating arsenic fate using integrated spectroscopic and microscopic techniques. Science of the Total Environment. 2018;**621**:1642-1651

[34] Pan H, Yang X, Chen H, Sarkar B, Bolan N, Shaheen SM, et al. Pristine and iron-engineered animal-and plant-derived biochars enhanced bacterial abundance and immobilized arsenic and lead in a contaminated soil. Science of the Total Environment. 2021;**763**:144218

[35] Poveda J, Martínez-Gómez Á, Fenoll C, Escobar C. The use of biochar for plant pathogen control. Phytopathology. 2021;**111**:1490-1499

[36] Kammann C, Ratering S, Eckhard C, Müller C. Biochar and hydrochar effects on greenhouse gas (carbon dioxide, nitrous oxide, and methane) fluxes from soils. Journal of Environmental Quality. 2012;**41**:1052-1066

[37] Laird DA, Fleming P, Davis DD, Horton R, Wang B, Karlen DL. Impact of biochar amendments on the quality of a typical Midwestern agricultural soil. Geoderma. 2010;**158**:443-449

[38] Lehmann J, Rillig M, Thies J, Masiello CA, Hockaday WC, Crowley D. Biochar effects on soil biota: A review. Soil Biology and Biochemistry. 2011;**43**:1812-1836

[39] Prendergast-Miller MT, Duvall M, Sohi SP. Biochar-root interactions are mediated by biochar nutrient content and impacts on soil nutrient availability. European Journal of Soil Science. 2014;**65**:173-185

[40] Zhang M, Riaz M, Zhang L, El-Desouki Z, Jiang C. Biochar induces changes to basic soil properties and bacterial communities of different soils to varying degrees at 25 mm rainfall: More effective on acidic soils. Frontiers in Microbiology. 2019;**10**:1321

[41] Ahluwalia O, Singh P, Bhatia R. A review on drought stress in plants: Implications, mitigation and the role of plant growth promoting rhizobacteria. Resources, Environment and Sustainability. 2021;**5**:100032

[42] Wang G, Ma Y, Chenia HY, Govinden R, Luo J, Ren G. Biochar-mediated control of *Phytophthora* blight of pepper is closely related to the improvement of the rhizosphere fungal community. Frontiers in Microbiology. 2020;**11**:1427

[43] Elmer W. Effect of leaf mold mulch, biochar, and earthworms on mycorrhizal colonization and yield of asparagus affected by *Fusarium* crown and root rot. Plant Disease. 2016;**100**:2507-2512

[44] Elmer WH, Pignatello JJ. Effect of biochar amendments on mycorrhizal associations and *Fusarium* crown and root rot of asparagus in replant soils. Plant Disease. 2011;**95**:960-966

[45] Agrios GN. Plant Pathology. USA: Elsevier; 2005

[46] Mishra RK, Mohanty K. A review of the next-generation biochar production

from waste biomass for material applications. Science of the Total Environment. 2023;**904**:167171

[47] Weber K, Quicker P. Properties of biochar. Fuel. 2018;**217**:240-261

[48] Mukherjee A, Zimmerman AR. Organic carbon and nutrient release from a range of laboratory-produced biochars and biochar–soil mixtures. Geoderma. 2013;**193**:122-130

[49] Li H, Tan Z. Preparation of high water-retaining biochar and its mechanism of alleviating drought stress in the soil and plant system. Biochar. 2021;**3**:579-590

[50] Rafiq MK, Bachmann RT, Rafiq MT, Shang Z, Joseph S, Long R. Influence of pyrolysis temperature on physico-chemical properties of corn stover (*Zea mays* L.) biochar and feasibility for carbon capture and energy balance. PLoS One. 2016;**11**:e0156894

[51] Tomczyk A, Sokołowska Z, Boguta P. Biochar physicochemical properties: Pyrolysis temperature and feedstock kind effects. Reviews in Environmental Science and Bio/Technology. 2020;**19**:191-215

[52] Enders A, Hanley K, Whitman T, Joseph S, Lehmann J. Characterization of biochars to evaluate recalcitrance and agronomic performance. Bioresource Technology. 2012;**114**:644-653

[53] Jindo K, Mizumoto H, Sawada Y, Sanchez-Monedero MA, Sonoki T. Physical and chemical characterization of biochars derived from different agricultural residues. Biogeosciences. 2014;**11**:6613-6621

[54] Singh B, Singh BP, Cowie AL. Characterisation and evaluation of biochars for their application as a soil amendment. Soil Research. 2010;**48**:516-525

[55] Lei O, Zhang R. Effects of biochars derived from different feedstocks and pyrolysis temperatures on soil physical and hydraulic properties. Journal of Soils and Sediments. 2013;**13**:1561-1572

[56] Asadi H, Ghorbani M, Rezaei-Rashti M, Abrishamkesh S, Amirahmadi E, Chengrong C, et al. Application of rice husk biochar for achieving sustainable agriculture and environment. Rice Science. 2021;**28**:325-343

[57] Domingues RR, Trugilho PF, Silva CA, ICND M, Melo LC, Magriotis ZM, et al. Properties of biochar derived from wood and high-nutrient biomasses with the aim of agronomic and environmental benefits. PLoS One. 2017;**12**:e0176884

[58] Duku MH, Gu S, Hagan EB. A comprehensive review of biomass resources and biofuels potential in Ghana. Renewable and Sustainable Energy Reviews. 2011;**15**:404-415

[59] Ghorbani M, Neugschwandtner RW, Konvalina P, Asadi H, Kopecký M, Amirahmadi E. Comparative effects of biochar and compost applications on water holding capacity and crop yield of rice under evaporation stress: A two-years field study. Paddy and Water Environment. 2023;**21**:47-58

[60] Liu Z, Jia M, Li Q, Lu S, Zhou D, Feng L, et al. Comparative analysis of the properties of biochars produced from different pecan feedstocks and pyrolysis temperatures. Industrial Crops and Products. 2023;**197**:116638

[61] Nwajiaku IM, Olanrewaju JS, Sato K, Tokunari T, Kitano S, Masunaga T. Change in nutrient

composition of biochar from rice husk and sugarcane bagasse at varying pyrolytic temperatures. International Journal of Recycling of Organic Waste in Agriculture. 2018;7:269-276

[62] Li S, Harris S, Anandhi A, Chen G. Predicting biochar properties and functions based on feedstock and pyrolysis temperature: A review and data syntheses. Journal of Cleaner Production. 2019;**215**:890-902

[63] Suga H, Hyakumachi M. Genomics of phytopathogenic Fusarium. In: Applied Mycology and Biotechnology. Volume 4: Fungal Genomics. Netherlands: Elsevier Science B.V.; 2004. pp. 161-189

[64] Rossman AY, Palm ME. Why are Phytophthora and other Oomycota not true fungi? Outlooks on Pest Management. 2006;**17**:217

[65] Nehra S, Gothwal RK, Varshney AK, Solanki PS, Chandra S, Meena P, et al. Chapter 19 - bio-management of Fusarium spp. associated with fruit crops. In: Sharma VK, Shah MP, Parmar S, Kumar A, editors. Fungi Bio-Prospects in Sustainable Agriculture, Environment and Nano-Technology. Academic Press; 2021. pp. 475-505. Available from: https://shop.elsevier.com/books/fungi-bio-prospects-in-sustainable-agriculture-environment-and-nano-technology/sharma/978-0-12-821734-4

[66] Akhter A, Hage-Ahmed K, Soja G, Steinkellner S. Compost and biochar alter mycorrhization, tomato root exudation, and development of *Fusarium oxysporum* f. sp. lycopersici. Frontiers in Plant Science. 2015;**6**:529

[67] Eo J, Park K-C, Kim M-H, Kwon S-I, Song Y-J. Effects of rice husk and rice husk biochar on root rot disease of ginseng (*Panax ginseng*) and on soil organisms. Biological Agriculture & Horticulture. 2018;**34**:27-39

[68] Wang Q, Fang W, Yan D, Han D, Li Y, Ouyang C, et al. The effects of biochar amendment on dimethyl disulfide emission and efficacy against soil-borne pests. Water, Air, & Soil Pollution. 2016;**227**:1-9

[69] Iacomino G, Idbella M, Laudonia S, Vinale F, Bonanomi G. The suppressive effects of biochar on above- and belowground plant pathogens and pests: A review. Plants. 2022;**11**:3144

[70] Akanmu A, Sobowale A, Abiala M, Olawuyi O, Odebode A. Efficacy of biochar in the management of *Fusarium verticillioides* Sacc. causing ear rot in *Zea mays* L. Biotechnology Reports. 2020;**26**:e00474

[71] Asif M, Haider MS, Akhter A. Impact of biochar on *Fusarium* wilt of cotton and the dynamics of soil microbial community. Sustainability. 2023;**15**:12936

[72] Liu Z, Zhou W, Sun Y, Peng Y, Niu J, Tan J, et al. Biochar and its coupling with microbial inoculants for suppressing plant diseases: A review. Applied Soil Ecology. 2023;**190**:105025

[73] Judelson HS, Blanco FA. The spores of *Phytophthora*: Weapons of the plant destroyer. Nature Reviews Microbiology. 2005;**3**:47-58

[74] Thines M. Oomycetes. Current Biology. 2018;**28**:R812-R813

[75] Lamour KH, Stam RJ, Julietta HE. The oomycete broad-host-range pathogen *Phytophthora capsici*. Molecular Plant Pathology. 2012;**13**:329-337

[76] Gadgil PD. Phytophthora Heveae, a Pathogen of Kauri. New Zealand: New Zealand Forest Service; 1974

[77] Volynchikova E, Kim KD. Biological control of oomycete *Soilborne* diseases caused by *Phytophthora capsici*, *Phytophthora infestans*, and *Phytophthora nicotianae* in Solanaceous crops. Mycobiology. 2022;**50**:269-293

[78] Zwart DC, Kim SH. Biochar amendment increases resistance to stem lesions caused by *Phytophthora* spp. in tree seedlings. HortScience. 2012;**47**:1736-1740

[79] Shoaf N, Hoagland L, Egel DS. Suppression of *Phytophthora* blight in sweet pepper depends on biochar amendment and soil type. HortScience. 2016;**51**:518-524

[80] Postma J, Clematis F, Nijhuis EH, Someus E. Efficacy of four phosphate-mobilizing bacteria applied with an animal bone charcoal formulation in controlling *Pythium aphanidermatum* and *Fusarium oxysporum* f.sp. radicis lycopersici in tomato. Biological Control. 2013;**67**:284-291

[81] Cruz DR, Leandro LFS, Munkvold GP. Effects of temperature and pH on *Fusarium oxysporum* and soybean seedling disease. Plant Disease. 2019;**103**:3234-3243

[82] Jayaraman S, Naorem AK, Lal R, Dalal RC, Sinha NK, Patra AK, et al. Disease-suppressive soils-beyond food production: A critical review. Journal of Soil Science and Plant Nutrition. 2021;**21**:1437-1465

[83] Orr R, Nelson PN. Impacts of soil abiotic attributes on *Fusarium* wilt, focusing on bananas. Applied Soil Ecology. 2018;**132**:20-33

[84] Segura RA, Stoorvogel JJ, Sandoval JA. The effect of soil properties on the relation between soil management and *Fusarium* wilt expression in Gros Michel bananas. Plant and Soil. 2022;**471**:89-100

[85] Mahala DM, Maheshwari HS, Yadav RK, Prabina BJ, Bharti A, Reddy KK, et al. Microbial transformation of nutrients in soil: An overview. In: Rhizosphere Microbes: Soil and Plant Functions. Singapore: Springer Nature; 2020. pp. 175-211

[86] Xie P, Li C-L, Shao B, Xu X-J, Chen X-D, Zhao L, et al. Simultaneous removal of carbon dioxide, sulfur dioxide and nitric oxide in a biofilter system: Optimization operating conditions, removal efficiency and bacterial community. Chemosphere. 2021;**276**:130084

[87] Naqvi S. Phytophthora diseases of citrus and management strategies. Annual Review of Plant Pathology (India: Scientific Publishers). 2003;**2**:239-270

[88] Loekito S, Afandi, Afandi A, Nishimura N, Koyama H, Senge M. Study on Soil Properties and Species Conformity of Phytophthora Species in a Pineapple Field. Pakistan: CABI; 2022

[89] Graber ER, Frenkel O, Jaiswal AK, Elad Y. How may biochar influence severity of disease caused by soilborne pathogens? Carbon Management. 2015;**5**:169-183

[90] Höper HC, Steinberg C, Alabouvette C. Involvement of clay type and pH in the mechanisms of soil suppressiveness to *Fusarium* wilt of flax. Soil Biology and Biochemistry. 1995;**27**:955-967

[91] Graber ER, Harel YM, Kolton M, Cytryn E, Silber A, David DR, et al. Biochar impact on development and productivity of pepper and tomato grown in fertigated soilless media. Plant and Soil. 2010;**337**:481-496

[92] Elad Y, David DR, Harel YM, Borenshtein M, Kalifa HB, Silber A, et al. Induction of systemic resistance in plants by biochar, a soil-applied carbon sequestering agent. Phytopathology. 2010;**100**:913-921

[93] Harel YM, Elad Y, Rav-David D, Borenstein M, Shulchani R, Lew B, et al. Biochar mediates systemic response of strawberry to foliar fungal pathogens. Plant and Soil. 2012;**357**:245-257

[94] Mehari ZH, Elad Y, Rav-David D, Graber ER, Harel YM. Induced systemic resistance in tomato (*Solanum lycopersicum*) against *Botrytis cinerea* by biochar amendment involves jasmonic acid signaling. Plant and Soil. 2015;**395**:31-44

[95] Graber ER, Elad Y. Biochar impact on plant resistance to disease. In: Ladygina N, Rineau F, editors. Biochar and Soil Biota. Boca Raton, FL: CRC Press; 2013. pp. 41-68

[96] Agrios G. Plant Pathology 5th Edition. Burlington, Ma. USA: Elsevier academic press; 2005. pp. 79-103

[97] Glaser B, Lehmann J, Zech W. Ameliorating physical and chemical properties of highly weathered soils in the tropics with charcoal–a review. Biology and Fertility of Soils. 2002;**35**:219-230

[98] Ahmad CA, Akhter A, Haider MS, Abbas MT, Hashem A, Avila-Quezada GD, et al. Demonstration of the synergistic effect of biochar and *Trichoderma harzianum* on the development of *Ralstonia solanacearum* in eggplant. Frontiers in Microbiology. 2024;**15**:1360703

[99] Atucha A, Litus G. Effect of biochar amendments on peach replant disease. HortScience. 2015;**50**:863-868

[100] Cobb RC, Haas SE, Kruskamp N, Dillon WW, Swiecki TJ, Rizzo DM, et al. The magnitude of regional-scale tree mortality caused by the invasive pathogen *Phytophthora ramorum*. Earth's Future. 2020;**8**:e2020EF001500

[101] Donald F, Purse BV, Green S. Investigating the role of restoration plantings in introducing disease—A case study using *Phytophthora*. Forests. 2021;**12**:764

[102] Green S, Cooke DE, Dunn M, Barwell L, Purse B, Chapman DS, et al. Phyto-threats: Addressing threats to UK forests and woodlands from *Phytophthora*. Forests. 2021;**12**(12):1617

[103] Wingfield M, Hammerbacher A, Ganley R, Steenkamp E, Gordon T, Wingfield B, et al. Pitch canker caused by *Fusarium circinatum*—A growing threat to pine plantations and forests worldwide. Australasian Plant Pathology. 2008;**37**:319-334